AF517840

ENCYCLOPAEDIA OF ENGINEERING

ENCYCLOPAEDIA
OF
ENGINEERING

VOL. 3
CIVIL ENGINEERING
Q-Z

By

A.M. Kharmawphlang
A. Khongphai

AKANSHA PUBLISHING HOUSE
NEW DELHI

AKANSHA PUBLISHING HOUSE
4649-B/21, Ansari Road
Darya Ganj, New Delhi-110 002
Email: ektabooks@yahoo.com
Ph.: 23263193/9811582579
Fax: 011-23263193

Encyclopaedia of Engineering

First Published 2006

ISBN 81-8370-048-9 (Set)

PRINTED IN INDIA

Published by M.P. Misra for Akansha Publishing House, New Delhi and Printed at Aryan Enterprise, Delhi.

PREFACE

Engineering is the application of science to the needs of humanity. This is accomplished through the application of knowledge, mathematics, and practical experience to the design of useful objects or processes.

Engineering involves the knowledge of mathematical and natural sciences (biological and physical) gained by study, experience, and practice, applied with judgement and creativity to develop ways to utilize the materials and forces of nature for the benefit of mankind.

Encyclopaedia of engineering contains engineering terms with clear and concise definitions. The encyclopaedia is highly illustrated and written in plain, non-technical language and extensive cross-referencing makes concepts more easily understood. Timings and Twigg focus on a wide range of engineering disciplines — engineering science, electrical and electronic engineering, workshop practices and mechanical engineering — and provide definitions that are relevant to all.

This vital reference offers a wealth of essential information in a convenient, quick-find format. Whether you're a professional, a student, a writer, or a general reader with an interest in science, there is no better or more authoritative way to stay up-to-speed with the current language of the various branches of engineering or gain an understanding of their key ideas and concepts.

Encyclopaedia of Engineering is written specifically for further education students by experienced authors who have taught and written extensively at this level.

Q

Quad. 1. A unit of energy equal to one quadrillion (1,000,000,000,000,000) BTUs. 2. Quadrangle. 3. Quadriphonic.

Quadrangle. 1. A four sided enclosure surrounded by buildings. 2. A quadrilateral. 3. One of the survey map sheets published by the USGS.

Quadrant. 1. A quarter of a circle. 2. An instrument for measuring altitudes. 3. A hardware device to fasten together the upper and lower leaves of a Dutch door.

Quadrel. A square tile; a quarrel.

Quadrilateral. A polygon of four sides.

Quadripartite. Having four parts; quadripartite vaulting is the commonest form of groined and also of ribbed vaulting.

Quadriphonic. Sound recorded or transmitted utilizing four transmission channels.

Quadrisect. To divide into 4, usually equal, parts

Quadruplex Cable. Four wire cable.

Quag. A marsh.

Quagmire. A soft boggy or marshy area that gives way under foot.

Quake. An earthquake.

Qualified Historical Building (Or Structure). Any structure or collection of structures deemed important to the history, architecture or culture of an area by an appropriate local or state governmental jurisdiction. This shall include structures of national, state or local historical registers or official inventories, such as the National Register of Historic Places, State Historical Landmarks, State Points of Historical Interest, and city or county registers or inventories of historical or architecturally significant sites and places or landmarks.

Quality. The degree of excellence of a thing.

Quality Control. All the activities undertaken to ensure adequate quality in manufactured products and in on-site construction.

Quantity. The size, extent, weight, amount, number, or volume of a thing.

Quantity Survey. Detailed listing of all materials, equipment, and services needed to complete a project.

Quantity Surveyor. A person who measures and prices building work; see Quantity Survey.

Quantum. A desired or allowed amount.

Quarrel. 1. A square or lozenge-shaped piece of material, especially a piece of glass, set diagonally, as in a latticed sash; also called a quarrel pane. 2. The opening in a sash prepared for such glazing.

Quarry. An open excavation for removing building stone, slate, or limestone; a rock pit.

Quarry Bed. The side of a piece of building stone that is parallel with the natural strata or veins.

Quarry-Faced. Stone having a rough face as if just split from the quarry.

Quarry Tile. A large, fired, clay floor tile, usually unglazed.

Quarter Bend. A drainage pipe fitting which makes a 90 degree angle.

Quarter Drop Match. See Set Match.

Quarter Hollow. A concave molding or cavetto of which the transverse section is an arc of about 90 degrees, the converse of a quarter round, or ovolo.

Quarter Round. Wood molding with a cross section in the shape of a quarter circle.

Quartersawed Lumber. Edge-grained lumber.

Quarter Width. The unit of yard measure (1/4) used in referring to carpet or loom widths; early European carpet was woven in widths of 27 inches or 3/4 yards; hence, 4/4 = 1-yard width or 3′; 12/4 width = 9′ and 16/4 width = 12′.

Quartz. Crystalline silica; a hard brittle mineral breaking with a glasslike fracture, and usually transparent to translucent, and colorless, or of a white, pink, and amethystine hue.

Quartzite. A strong sandstone cemented by quartz, about 98 percent silica.

Quasi Contract. Application of contract rules of law to parties that do not have a contractual relationship.

Quatrefoil. A four cusped circular element in Gothic tracery.

Quay. A structure built along the bank of a waterway for use as a landing place.

Queen. 1. A half brick made by cutting a whole brick in two lengthwise as for a corner in a soldier course. 2. A half brick used to prevent vertical joints falling above one another.

Queen Anne. The architecture and furniture style of England, under Dutch influence, during the short reign of Anne, 1702-1714.

Queen Closer. A brick cut in half lengthwise.

Queen Truss. A truss framed with queen posts.

Quenching. 1. Cooling of steel in cold water to increase its hardness. 2. Rapid cooling of metal in a heat treating process.

Quick Assets. 1. Assets easily converted to cash. 2. Current assets without inventory.

Quick Burst. The inter-pressure required to burst a pipe or fitting due to an internal pressure build-up, usually within 60 to 90 seconds.

Quick-Connect Coupling. A device which permits easy and fast connecting of two fluid lines.

Quick Drying. A material with a relatively short drying time.

Quick Fix. An expedient often inadequate solution to a problem; a Band-Aid Approach.

Quicklime. The first solid product CaO that is obtained by calcining limestone and that develops great heat and becomes crumbly when treated with water; lime; before it can be used in construction, quicklime must be slaked in water and aged for at least 2 weeks.

Quicksand. 1. Fine sand or silt that is prevented from settling firmly together by upward movement of ground water. 2. Any wet inorganic soil so unsubstantial that it will not support any load.

Quick Set. 1. See Flash Set. 2. In gypsum wallboard joints, a joint compound that chemically hydrates prior to drying; also called Fast Set.

Quicksilver. Mercury.

Quiet Title Action. A lawsuit brought to remove a claim on the title of real property.

Quincunx. An arrangement of five objects in which four are at the corners and one is at the center at the crossed diagonals, in the manner of a dice or a five playing card, used in planting trees.

Quintefoil. A five cusped element in Gothic tracery.

Quirk. A groove separating two parts of a molding.

Quitclaim Deed. A form of deed in the nature of a release containing both words of grant and of release; it does not provide a guarantee or warranty of title.

Quoin. Cornerstones forming the external angle of a building; bricks or block masonry that project out from the walls at the corners of a building for ornamental effect.

Quonset. A prefabricated metal building with a semicylindrical corrugated roof.

R

R&D. Research and Development.

R. 1. Radius of gyration. 2. Electrical Resistance, in Ohms.

R-12. Dichlorodifluromethane.

RA. 1.Registered Architect. 2. Return Air.

Rabbet. A corner cut out along the corner edge of a piece of wood.

Raceway. Any channel courses supporting and protecting electrical conductors, including conduits, wireways, surface metal raceway, cable trays, floor and ceiling raceways, busways, and cable bus.

Rack. 1. A framework or stand to support materials or objects. 2. A toothed bar that meshes with a gear wheel as in a Rack and Pinion.

Rack and Pinion. A mechanical device consisting of a toothed metal bar (the rack) and a small meshing gear wheel (the pinion).

Racking. 1. Lateral stresses exerted on an assembly. 2. In bricklaying, the method of laying the end of a wall in a series of steps so when work is resumed at a later date, the bond will continue with the old construction easily.

Rack-Type Dishwasher. A dishwasher that accepts dishes which are loaded on metal or plastic racks.

Rad. Radius.

Radial. Moving along lines diverging from a center, like spokes.

Radial Commutator. Electrical contact surfaces on a rotor which is perpendicular or at right angles to the shaft center line.

Radial Wall Form. A temporary structure or mold for the support of a curved concrete wall while the concrete is curing.

Radian. A unit of circular angle, equal to an angle at the center of a circle the arc of which is equal in length to the radius; one radian equals 57.2958 degrees; one degree equals 0.017454 radians.

Radiant Heat Transfer. Radiant heat transfer occurs when there is a large difference between the temperatures of two surfaces that are exposed to each other, but are not touching.

Radiant Heat. Heat transmitted by radiation.

Radiant Heating. A heating system in which warm or hot surfaces are used to radiate heat into the space to be conditioned.

Radiant Panels. Panels with integral passages for the flow of warm air or liquids; heat is transferred to a room by thermal radiation.

Radiation. 1. Heat transformed by electromagnetic waves. 2. A net exchange of radiant energy between two bodies or objects, across an open space.

Radiation Unit. A determinate quantity adopted as a standard of measurement of radiant heating.

Radiator. A heating device that transfers heat from the heated fluid within to the air outside.

Radio Frequency Curing. Of wood adhesives, curing of bondlines by the application of radio frequency energy; also called High Frequency Curing.

Radiology Equipment. Apparatus or devices that use high energy radiation in medicine.

Radius. A straight line from the center to the circumference of a circle or sphere; one- half of the diameter.

Radius of Gyration. A characteristic of the cross-section of the member used in the determination of its structural characteristics.

Radon. A heavy radioactive gaseous element formed by disintegration of radium; it seeps out of the earth and accumulates under buildings, in some cases causing cancer in humans; easily removed by passive ventilation.

Raft. A footing or foundation, usually a large thick concrete mat.

Rafter. A framing member that runs up and down the slope of a pitched roof; the parallel beams that slope from the ridge of a roof to the eaves and make up the main body of the roof framework.

Rafter Anchor. A bolt or metal fastening device which attaches the rafters wall plate.

Rafter, Common. See Common Rafter.

Rafter, Cripple. See Cripple Rafter.

Rafter, Hip. See Hip Rafter.

Rafter, Jack. See Jack Rafter.

Rafter Tie. A wood or steel horizontal tension member connecting two opposite rafters to resist their spreading.

Rafter, Valley. See Valley Rafter.

Raggle Block. A concrete masonry unit with an integral flashing reglet.

Raggle. A groove cut in masonry; any groove provided to receive roofing or flashing.

Rail. (1) A horizontal bar or timber extending from one post or support to another, such as a guard or barrier in a fence or staircase. (2) The horizontal members of the framework of a sash, door, blind, or any paneled assembly. 3. A horizontal support member to support a curtain.

Rail, Chair. See Chair Rail.

Railing. 1. An open fence or guard for safety, made of rails and posts. 2. A banding in cabinetwork. 3. On plywood, the solid wood band around one or more edges.

Railing Expansion Joint. Control joint in metal railing to allow differential movement of railing components from thermal expansion and contraction.

Railing Fitting. Elbows, crosses, tees, caps, flanges made of metal, slip type, threaded, or flush welded to railing components to facilitate transitions and terminations.

Railing Flange. Flanges made of metal, slip type, threaded, or flush welded to railing components to facilitate attachment to mounting surfaces.

Railing, Metal. See Metal Railing.

Railing, Pipe. See Pipe Railing.

Railing, Structural Tubing. See Structural Tubing Railing.

Railroad Crossing. An grade level intersection at which railroad tracks meet vehicular, pedestrian or other kinds of traffic.

Railroad Tie. A timber support for railroad tracks.

Railroad Work. Construction or repair performed on railroad tracks.

Rail Standard, Steel. See Steel Rail Standard.

Rail Transit Boarding Platform. A horizontal, generally level surface, whether raised above, recessed below, or level with a transit rail, from which persons embark/disembark a fixed rail vehicle.

Rain Spots. Defects on paint film caused by rain.

Raintight. So constructed or protected that exposure to a beating rain will not result in the entrance of water.

Raintight Enclosure. A building, a component of a building, or a structure that is impervious to the outside elements such as rain or snow.

Raintight Time Clock. A weatherproof time recording clock that has been manufactured to withstand the outside elements, such as rain or snow.

Raised Grain. A roughened condition of the surface of dressed lumber in which the hard latewood is raised above the softer earlywood but not torn loose from it.

Rake. 1. A hand implement with spaced teeth for gathering or loosening material like grass, gravel, or earth. 2.The sloping edge of a pitched roof. 3. The trim of a building extending in an oblique line, as rake dado or molding. 4. The end of a wall that slopes or racks back.

Raked Joint. In brick masonry, a type of joint which has the mortar raked out to a specified depth while the mortar is still green.

Rake Molding. The cornice on the gable edge of a pitch roof, the members of which are made to fit those of the molding of the horizontal eaves.

Raking Bond. Bricks that are laid in a zigzag fashion.

RAM. Random Access Memory.

Ram Air. Air forced through the condenser due to the rapid movement of the vehicle along the highway.

Ramp. A sloping passageway.

Ranch Molding. An architectural style of molding that is installed as finish work or ornamentation; a type or style of trim which is gently curved and devoid of ornate design.

Random Access Memory. A memory in a computer system that is used for temporary storage of data and allows that data to be accessed at random and to be changed; compare Read Only Memory.

Random Rubble. Masonry wall built of unsquared or rudely squared stones, irregular in size and shape.

Random Width Flooring. Flooring materials of wood with varying widths, often called a plank floor.

Range. 1. Pressure or temperature settings of a control. 2. Change within limits. 3. A Kitchen Range.

Range, Electric. See Electric Range.

Range, Firing. See Firing Range, 2.

Range Hood. An exhaust system installed over a stove to remove fumes by means of a motorized fan and a duct to outside air.

Range, Kitchen. See Kitchen Range.

Rankine. A scale for registering temperature in which 0 degrees represents absolute zero; Rankine corresponds to the Fahrenheit scale, 32° F = 491.67° Rankine.

Rapid Curing (RC) Asphalt. Cutback asphalt composed of asphalt cement and a naphtha or gasoline-type diluent of high volatility.

Rapid Sand Filter. A swimming pool filter designed to be used with sand as the filter media and for flows not to exceed 3 GPM per square foot in commercial pools and 5 GPM in residential pools.

Rapid Start Lamp. A fluorescent lamp operated by a ballast that provides a low-voltage winding to preheat electrodes and initiate the arc without a starting switch or the application of high voltage.

Rare Gas. See Noble Gas.

Rasp. A coarse file with cutting points instead of lines.

Ratchet. A mechanical device, consisting of a toothed bar and pawl, that allows motion in one direction only.

Rate of Flow. Volume flow per unit of time.

Rate of Growth. The rate at which a tree has increased its amount of wood; this is measured radially in the trunk or in lumber cut from the trunk; the rate is determined from the number of annual growth rings per inch.

Rational Number. Expressible as a ratio of whole numbers, that is, not involving roots or non-terminating decimals; involving only addition, subtraction, multiplication and division and only a finite number of times.

Ratproofing. Various measures taken to keep rodents out of the interior of a building, particularly food preparation or storage areas, consisting mainly of steel mesh barriers above and below grade and packing of oversized pipe and conduit holes with cement mortar or steel wool.

Ravelling. In asphalt paving, the progressive separation of aggregate particles from the surface downward or from the edges inward; ravelling is caused by lack of compaction, construction of a thin lift during cold weather, dirty or disintegrating aggregate, too little asphalt in the mix, or overheating of the asphalt mix.

Raw. Raw linseed oil.

Raw Glaze. A ceramic glaze compounded primarily from raw constituents; it contains no pre-fused materials.

Raw Linseed Oil. The crude product processed from flaxseed, usually without much further treatment.

Raw Oil. Oil as received from the press or separated from the solvent in the solvent extraction process.

Raw Sienna. 1. An earthy substance containing oxides of iron and usually of manganese. 2. Brownish yellow pigment, used in paint. 3. See Sienna.

Raw Umber. 1. A brown earthy substance containing oxides of iron and manganese. 2. A pigment, darker than ochre and sienna, used in paint.

Rayleigh Wave. Forward and vertical vibration of seismic surface wave; see Seismic Wave.

Rays, Wood. See Wood Rays.

RBM. Reinforced Brick Masonry.

RC Asphalt. See Rapid Curing Asphalt.

RCP. Reinforced Concrete Pipe.

RCSSB. Red Cedar Shake and Shingle Bureau.

RDF. Refuse Driven Fuel.

Reaching. In using a spray gun, extending a spray stroke too far.

Reactance. That part of the impedance of an alternating current circuit due to capacitance or inductance or both.

Reaction. 1. In structures, the response of the structure to the loads. 2. The force that acts at a support of a structure.

Reaction Wood. Wood with more or less distinctive anatomical characters, formed typically in parts of leaning or crooked stems and in branches; in hardwoods this consists of tension wood and in softwoods of compression wood.

Reactive Aggregate. See Alkali-Aggregate Reaction.

Read Only Memory. A computer memory read only at high speeds but not capable of being changed; compare Random Access Memory.

Ready-Mixed. 1. Any material that is mixed prior to delivery to the jobsite. 2. Plaster which is mixed at the mill with mineral aggregate and other ingredients which control time of set; generally used in conjunction with gypsum plasters; also called mill-mixed, pre-mixed.

Ready-Mixed Compound. Factory mixed gypsumboard joint compound in ready-to-use form.

Ready-Mixed Concrete. Concrete manufactured at batch plants and delivered by truck to the job site in a plastic state.

Real Estate. Land that is subject to ownership, with its permanent improvements and appurtenances; also called Real Property.

Real Estate Broker. A broker who deals in real property.

Real Property. See Real Estate.

Realtor. A real estate agent who is a member of the National Association of Realtors.

Realty. Real property; real estate.

Reamer. A small hand tool for enlarging or shaping holes in wood, metal, or other materials.

Reaming. Removing the burr from inside of a pipe.

Rear Yard. The space between a building and the rear property line.

Reasonable Use. A legal doctrine that allocates water rights.

Rebar. Steel reinforcing bar.

Rebate. A continuous rectangular notch cut out of the corner of a piece of wood; a rabbet.

Receiver. A person appointed by the court to administer property in foreclosure or a business in reorganization.

Receiver-Drier. Cylinder (container) in a refrigerating system for storing liquid refrigerant and which also holds a quantity of desiccant.

Receiver Heating Element. Electrical resistance mounted in or around liquid receiver. It is used to maintain head pressures when ambient temperature is low.

Receptacle. An electric outlet.

Receptacle, Cable. See Cable Receptacle.

Receptacle, Combination. See Combination Receptacle.

Receptor. 1. A metallic or nonmetallic waterproof base for a shower stall. 2. A plumbing fixture, such as a floor sink, that receives the discharge from indirect waste piping.

Recess. A shallow depth or space in a wall; an alcove or niche.

Recessed. Set into a niche or recess in a wall, floor, or ceiling surface.

Recessed Edge. In gypsum board, a sharply reduced caliper along the factory edge; has been replaced by the more gradually reduced caliper of the Tapered Edge. See Taper.

Recessed Light Fixture. A light fixture installed in a suspended ceiling or a recess in a plaster or gypsum board ceiling.

Recess Filler Strip. A strip of material inserted in a fissure to make flush with the surrounding surface.

Reciprocal. 1. Inversely related; opposite; mutually corresponding. 2. One of a pair of numbers whose product is one (8 and 1/8 or 3/4 and 4/3).

Reciprocating. Back and forth motion in a straight line.

Reciprocating Chiller. A compressor (single-acting) using pistons that are driven by a connecting rod from a crankshaft.

Reciprocating Compressor. Compressor which uses a piston and cylinder mechanism to provide pumping action.

Reciprocating Single Piston Pump. A pump having a single reciprocating (moving up and down or back and forth) piston.

Recirculating Overflow. See Surface Skimmer.

Recirculation System. In swimming pool piping, the interconnected system traversed by the recirculated water from the pool until it is returned to the pool.

Recision. The cancellation of a contract.

Reclamation Tank. See Separation Tank.

Recoat Time. In painting, the time interval required between application of successive coats.

Reconstruction. The state of being constructed again.

Reconveyance Deed. A deed executed by the trustee under a trust deed after the underlying note has been satisfied.

Record Drawings. A set of drawings prepared by the general contractor, which includes any revisions in the working drawings and specifications during construction, indicating how the project was actually constructed; sometimes called As-Built Drawings or As-Builts.

Recording. The process of placing a document on public record by filing in the office of the county recorder; recording a document constitutes constructive notice.

Recording Ammeter. Electrical instrument which uses a pen to record amount of current flow on a moving paper chart.

Recording Thermometer. Temperature measuring instrument which has a pen marking a moving chart.

Recording Voltmeter. A device which records the continuity, power, grounding, voltage and resistance of electrical circuits.

Recourse. 1. The legal right to hold someone liable for payment such as the maker of a negotiable instrument. 2. The provision of some mortgages or trust deeds to hold the borrower personally responsible for the deficit when foreclosure sale proceeds are insufficient to cover the indebtedness.

Recreational Courts. A quadrangular space walled or marked off for playing games with a ball, such as basketball or tennis.

Recreational Facilities. Establishments and structures created for sporting and entertainment purposes, such as courts and pools.

Recreational Use Statutes. Laws that have been adopted in most states to protect property owners from liability to persons who are permitted to use the property for recreational purposes.

Rectangle. A four sided plane figure whose angles are equal; a parallelogram.

Rectifier. A device for converting alternating current to direct current.

Rectilinear. Characterized by straight lines; perpendicular.

Redan. A part of a medieval fortification consisting of two walls or parapets projecting at an acute angle and open to the rear.

Redemption. The repurchase of one's property after it has been sold in foreclosure; this right to redeem lapses after a specified time.

Red Fir. Douglas Fir.

Red Label Goods. Flammable or explosive materials with flash points below 80° F. (26.7° C.).

Red Lead. A compound formed by roasting lead or litharge; it used extensively in paints for protecting iron and steel against corrosion.

Red Ochre. An earthy pigment made from impure iron ore, used in paint.

Reducer. 1. A pipe fitting which connects pipes of different sizes. 2. A tile trim unit used to reduce the radius of a bullnose or a cove to another radius or to a square. 3. Volatile ingredients used to thin or lower viscosity of a finishing material.

Reducing Female Adapter. A pipe fitting which is threaded on the inside of at least one end to receive a different size pipe, to make a connection.

Reducing Flame. An oxygen-fuel gas flame wherein the portion used has a reducing effect; same as Carbonizing Flame.

Reducing Tee. A pipe fitting which joins three pipes of different sizes at 90 degree angles.

Redundant Member. Any member of a truss not required for truss stability.

Redwood. A coniferous timber tree with red heartwood and yellow sapwood.

Reel. A round device upon which a hose or wire is wound for easy use and storage.

Re-Entrainment. Situation that occurs when the air being exhausted from a building is immediately brought back into the system through the air intake and other openings in the building envelope.

Reentrant Angle. An angle pointing inward; the opposite of a salient angle.

Re-Entrant Corner. An inside corner of a surface, producing stress concentrations in the roofing or waterproofing membrane.

Refectory. A dining hall.

Refinancing. The reorganization of the loans on a property usually by finding new loans at lower interest rates or larger amounts and paying off the old loans.

Refined Shellac. A grade of orange or white shellac from which the wax has been removed.

Reflectance. The ratio of light reflected by a surface to the incident light falling on the surface.

Reflected Ceiling Plan. An architectural drawing of a ceiling seen as though it is reflected from the floor.

Reflection. The return of light and sound waves from a surface.

Reflection Cracks. Cracks in asphalt overlays that reflect the crack pattern in the pavement structure underneath; caused by vertical or horizontal movement in the pavement beneath the overlay, brought on by expansion and contraction with temperature or moisture changes.

Reflection Pool. An ornamental, usually shallow, pool of water designed to reflect an important architectural element beyond.

Reflective Coated Glass. Glass onto which a thin layer of metal or metal oxide has been deposited to reflect light and/or heat.

Reflectorized Sign. A sign made with highly reflective material.

Reflective Paint. Paint that has been formulated to reflect light away.

Reflector Lamp. Flood lamp or spot with bulb coated with a reflecting surface.

Reflector Lighting. A lighting fixture with a polished surface to redirect its light.

Refraction. The bending of a light ray as it passes obliquely through a material.

Refractive Index. Ratio of velocity of light in a certain medium compared with its velocity in air under same conditions.

Refractories. Heat-resistant non-metallic ceramic materials.

Refractory. Capable of enduring high temperature.

Refractory Concrete. Concrete having refractory properties, and suitable for use at high temperatures (generally about 315 to 1315° C), in which the binding agent is a hydraulic cement.

Refrigerant Charge. Quantity of refrigerant in a system.

Refrigerant. Substance used in refrigerating mechanism; it absorbs heat in evaporator by change of state from a liquid to a gas, and releases its heat in a condenser as the substance returns from gaseous state back to a liquid state.

Refrigerant Control. Device which meters flow of refrigerant between two areas of a refrigerating system; can be a capillary tube, expansion valves, or high and low-side valves; maintains pressure difference between high-pressure and low-pressure side of the mechanical refrigerating system while unit is running.

Refrigerant Cylinder. Cylinder in which refrigerant is stored and dispensed; color code painted on cylinder indicates kind of refrigerant.

Refrigerated Case. A storage unit using refrigeration to keep food or materials cool.

Refrigerated Truck. Commercial vehicle equipped to maintain below-ambient temperatures in its storage compartment

Refrigeration. The cooling of a material or space.

Refrigeration Oil. Specially prepared oil used in refrigerator mechanism which circulates, to some extent, with refrigerant.

Refrigeration Tubing. The hollow, cylindrical piping which holds the refrigerant fluid in a refrigeration system.

Refrigerator, Blood. See Blood Refrigerator.

Refuse Driven Fuel (RDF). A method of recovering heat, in an incinerator, from solid waste.

Refuse Hopper. A funnel-shaped receptacle having a device for releasing waste materials through a pipe.

Regional Shopping Center. A large shopping center serving a region.

Register. Combination grille and damper assembly covering an air opening or the end of an air duct.

Registered Architect. An architect who is registered with the state board of architectural examiners of one of the states; a licensed architect.

Register, Wall. Awall mounted grille or damper for the passage of conditioned air.

Reglet. A horizontal or sloping slot in masonry construction, or a sheetmetal slot in plastered wood construction, to accept the upper edge of counterflashing.

Regular Core Gypsumboard. Gypsumboard for general construction purposes without special core additives.

Regulations. Laws that are enacted by a public administrative agency, rather than by an elected legislative body.

Regulator. A device for controlling the delivery of gas at some substantially constant pressure regardless of variation in the higher pressure at the source.

Rehabilitation. The remodeling of a building to render it suitable for modern uses.

Rehydrate. To provide water to something that is dehydrated.

Reinforced Brick Masonry (RBM). Brickwork into which steel bars have been embedded in the mortar to impart tensile strength to the construction.

Reinforced Concrete Pipe. Factory-made precast piping

Reinforced Concrete. Concrete containing adequate reinforcement (pre-stressed or not prestressed) and designed on the assumption that concrete and steel act together in resisting forces; concrete work into which steel bars have been embedded to impart tensile strength to the member.

Reinforced Grouted Masonry. Wall construction consisting of brick or block that is grouted solid throughout its entire height and has both vertical and horizontal reinforcing.

Reinforced Hollow Unit Masonry. Wall construction consisting of hollow masonry units in which part of all of the cells are continuously grouted and has both horizontal and vertical reinforcing.

Reinforced Masonry. Unit masonry in which reinforcement is embedded in such a manner that the two materials act together in resisting forces.

Reinforced Membrane. A roofing or waterproofing membrane reinforced with felts, mats, fabrics or chopped fibers.

Reinforced Plastics. Plastics reinforced with glass or carbon fibers.

Reinforcement. 1. The action or state of strengthening by additional assistance, material, or support. 2. The action or state of strengthening or increasing by fresh additions. 3. Steel bars or wires embedded in concrete and located in such a manner that the metal and the concrete act together to resist loads.

Reinforcement, Mesh. See Welded Wire Fabric.

Reinforcing. 1. The act of making stronger. 2. A structure or member that is used to add strength. 3. Steel rods or metal fabric placed in concrete slabs, beams, or columns to increase their strength.

Reinforcing Accessory. Items used to install reinforcing in concrete, including, but not limited to, chairs, couplings, and tie wire.

Reinforcing Bar. A manufactured, usually deformed, steel bar, used in concrete and masonry construction to resist tensile stresses.

Reinforcing, Masonry. See Masonry Reinforcing.

REL. Recommended Exposure Limits; recommendations made by NIOSH.

Relative Humidity. The ratio of the amount of water vapor actually present in the air to the amount present when the air is saturated with water vapor at the same temperature, expressed as a percentage.

Relay. An electromagnetic mechanism, moved by a small electrical current in a control circuit, which operates a valve or switch in an operating circuit.

Relay, AC. See AC Relay.

Release Agent. Material used to prevent bonding of concrete to a form surface.

Release Clause. A provision in a mortgage or trust deed which allows a portion of the encumbered property to be released from the mortgage upon the payment of a specified sum of money to the lender.

Reliction. An increase in land area caused by the gradual recession of sea or river water.

Relief. 1. Damages or court orders awarded in litigation. 2. Ornamented prominence of parts of figures above a plane surface.

Relief Joint. See Control Joint and Expansion Joint.

Relief Valve. A pressure safety valve installed on water heaters, hot water and boiler tanks to relieve pressure when it exceeds a preset level.

Relief Vent. In a DWV system, a vent coming from a vent stack and connecting to a horizontal branch at a point between the first fixture connection on that branch and the soil stack or waste stack; its purpose is to allow the circulation of air between the vent stack and the soil or waste stack.

Relief Ventilator. An auxiliary vent whose main purpose is to provide supplementary circulation of air.

Relieving Arch. See Discharging Arch.

Reline Pipe. To install new linings in pipes; usually includes the cleaning of built-up scale or debris from the existing pipe and relining with a compatible material.

Reluctance. A force working against the passage of magnetic lines of force (flux) through a magnetic substance.

Remainder. The estate in real property that remains after a life estate terminates upon the death of the holder.

Remainderman. The person entitled to the estate that remains after termination of a life estate.

Remaining Economic Life. The time period from the present to the point where a real property improvement becomes valueless.

Remote Control Circuit. Any electrical circuit which controls any other circuit through a relay or an equivalent device.

Remote Power Element Control. Device with sensing element located apart from the operating mechanism it controls.

Remote System. Refrigerating system in which condensing unit is away from space to be cooled.

Removable Railing Section. Section of a railing designed and constructed to be removed easily to facilitate movement through the railing.

Removable Window Well Cover. Removable steel bar or expanded metal cover for a window well.

Remove. The act or process of demolishing, dismantling, and carrying away an old fitting or component.

Removers. Compositions designed to soften old varnish or paint coats so that they may be easily removed by scraping or washing.

Renaissance. 1.Rebirth; revival. 2. The art and architecture developed during the period of 14-17th century; the transition between medieval and modern times, marked by the revival of classical styles and the beginnings of modern science.

Renaissance Cities. Formal plans of a symmetric nature; hierarchical articulation of spaces between more prominent and less prominent parts of the cities, such as palaces, monuments, and other focal points.

Rendering. 1. An architectural drawing, often in color and in perspective. 2. Application of plaster directly to a masonry wall.

Rental Value. A measure of damages for deprivation of the use of property.

Repair. The reconstruction or renewal of any part of an existing building for the purpose of its maintenance.

Repeat. In carpet, fabric, and wallpaper, the distance from a point in a pattern figure to the same point where it occurs again, measuring lengthwise of the material.

Replacement Reserves. A portion of the annual income set aside for replacement of wasting assets, such as roofing, mechanical and electrical equipment, and other impermanent building components.

Replevin. A legal process whereby personal property that has been unlawfully taken from one may be recovered.

Replication. A copy or reproduction.

Reprographics. Systems for reproducing graphic materials, including architectural and engineering drawings and diagrams.

Repulsion-Start Induction Motor. An electric motor type which has an electrical winding on the rotor for starting purposes.

Required Strength. The strength of a member or cross section required to resist factored loads or related internal moments and forces.

Reredos. 1. An ornamental wall behind an altar in a church. 2. The back of a fireplace or open hearth.

Reproofing. The practice of applying new roofing materials over existing roofing materials.

Resawing. Sawing lumber again after the first sawing; specifically, sawing into boards or dimension lumber.

Rescission. A legal remedy for a material breach of contract under which the law pretends that a contract never existed and attempts to put the parties in the positions they occupied before the contract was executed; a cancellation.

Reserves. Portions of current income set aside and accumulated at interest to pay for future maintenance, repairs, and replacement of equipment.

Residence. A building for human habitation; a dwelling.

Residential Carpet. A grade or style of carpet installed in low foot-traffic applications.

Residential Equipment. An appliance or piece of equipment found in the home, that is used by the homeowner.

Residential Occupancy. A building or any portion of a building containing one or more dwelling units.

Residual. 1. The remainder or residue after a chemical reaction or a transaction. 2. Free acting disinfectant remaining in poolwater after treating and breaking down pollution.

Residual, Building. See Building Residual.

Residual, Land. See Land Residual.

Residual Stress. Stress remaining in a structure or member as a result of thermal or mechanical treatment or both.

Resilience. 1. Ability of a material to resume its original size and shape after deformation, such as stretching, twisting, compression, or indentation. 2. The measurement of the absorption of dynamic energy by a structure without permanent deformation or fracture. 3. The property whereby a strained body gives up its stored energy on the removal of the deforming force. 4. The ability of a carpet fabric or padding to spring back to its original shape or thickness after being crushed or walked upon.

Resilient Channel. A metal furring member designed to absorb sound or noise impact which strikes the surfacing membrane.

Resilient Flooring. Flooring materials such as asphalt, vinyl, linoleum, rubber, cork, and similar resilient materials; available in tile or sheet form.

Resin. A sticky material obtained from the sap of certain trees and plants (natural resin) or made synthetically from coal-tar products and other organic substances (synthetic resin); resins are widely used in making varnishes and paints.

Resin Hardness. Method of indicating hardness of resins; usually from No. 1 (hardest) to No. 6.

Resistance. 1. An opposing or retardant force. 2. An opposition to flow or movement such as friction. 3. The property of hindering the conduction of electricity or heat.

Resistance or R-Value. The tendency of a material to retard the flow of heat.

Resistivity (r). The property of hindering the conduction of heat or electricity; the reciprocal of conductivity, 1/k.

Resistor. Electrical device having resistance to the passage of electrical current.

Resist Printing. Placing a dye-resist agent on carpet prior to piece dyeing so that the pile will absorb color according to a predetermined design.

Res Judicata. The doctrine that courts will not re-litigate the same issues between the same parties.

Resolving Forces. Replacing a force or forces with two or more other forces that yield the same effect on a structure as the original forces.

Resonance. Movement produced in an element as a result of movement added to its natural period of movement, which is the same.

Resorcinol Glue. A glue that is high in both wet and dry strength and resistant to high temperatures; used for gluing lumber or assembly joints that must withstand severe service conditions.

Respirator. A device for maintaining artificial respiration.

Respirator Mask. A device worn over the mouth to aid in breathing.

Respondiat Superior. The doctrine that a principal is liable for the acts of an agent, and an employer for the acts of an employee.

Response. Effect produced on a structure by earthquake ground motion.

Restaurant. A room or building for preparing and serving meals to the public.

Restoration. The act of restoring to an original condition; the act or process of bringing a structure back to its former position or condition.

Restretch. A term applied to the remedial steps necessary for the correction of improperly laid carpet resulting from application of wrong stretching techniques, carpet defects, or undetermined causes.

Restrictor. A device for producing a deliberate pressure drop or resistance in a line by reducing the cross-sectional flow area.

Restroom. A room in a public place equipped with toilets, urinals, and lavatories, usually segregated by gender.

Resultant. A force that will produce the same effect as two or more combined forces.

Retainage. Amount of construction sum held back by the owner from the contractor to be paid after construction is satisfactorily completed; also called Retention.

Retained Earnings. The accumulated earnings that are retained in a corporation since it was founded less the total of dividends declared to the stockholders.

Retaining Wall. A wall that is designed to resist the lateral pressures of retained soil; a wall that holds back a hillside or is backfilled to create a level surface.

Retardant, Fire. See Fire Retardant.

Retardation. Delaying the hardening or strength gain of fresh concrete, mortar or grout.

Retarder. 1. A concrete admixture used to slow down the natural curing process. 2. An admixture used to delay the setting action of plaster; generally used only with gypsum plasters or finish coat plaster containing calcined gypsum gauging. 3. Slow drying solvents or extenders added to lacquer to delay drying of the lacquer

Retemper. Addition of water to portland cement plaster, mortar, or concrete after mixing but before setting process has started.

Retention. Retainage.

Retrofitting. To install new materials or equipment in an existing building.

Return. 1. Any surface turned back from the face of a principal surface. 2. The ending of a small splash wall or a wainscot at right angle to the major wall. 3. The continuation of a molding or finish of any kind in a different direction. 4. In HVAC, a term for the return-air duct of a forced air heating/cooling system. 5. The continuation in a different direction of the face of a building or any member.

Return Air Duct. The duct by which air is returned from a room or space to the heating or cooling system to be reconditioned and re-used.

Return Lines. Pipework returning water to be recooled or reheated. 2. Pipework returning filtered water to the pool.

Return Off-Set. See Jumpover.

Return on Investment. The percentage of earnings on the total assets of a corporation.

Return Period of Earthquakes. The time period in which the probability that an earthquake of a certain magnitude will reoccur.

Return Piping. That part of the piping between the filter and the pool, spa or hot tub through which passes the filtered water.

Rev. 1. Revised. 2. Revision. 3. Reverse. 4. Reversed.

Reveal. The side of an opening, as a window or door jamb.

Revenues. The gross income of a business.

Reverberation. The persistence or echoing of previously generated sound caused by reflection of acoustic waves from the surfaces of enclosed spaces; the support of sound by successive reflections.

Reverberation Time. The time is takes sound to become inaudible; a function of a space, people, furnishings, and the absorptivity of the surfaces.

Reverse Coloring. The changing of yarn frames in Jacquard weaves to cause the interchanging of ground and top colors, according to customers' preferences.

Reverse Cycle Defrost. 1. Method of heating evaporator for defrosting purposes. 2. Valves to move hot gas from compressor into evaporator.

Reverse Engineering. The study and analysis of a product in order to gain information about its design, construction, and use, for designing a new improved version.

Reverse Thrust Fault. A geological fault under compression where the overlying block moves up the dip of the fault plane.

Reversible Lock. A lock that will function on doors of either hand, sometimes with slight adjustments.

Reversing Valve. Device used to reverse direction of the refrigerant flow depending upon whether heating or cooling is desired.

Reversionary Clause. A provision in a deed that provides for the property to revert to the grantor in the event that any of the restrictions are violated.

Revibration. The delayed vibration of concrete that has already been placed and consolidated; most effective when done at the latest time a running vibrator will sink of its own weight into the concrete and again make it plastic.

Revolving Door. Typically a four panel door attached at 90 degrees to each other that turns on a center axis; some are 3 panel attached at 120 degrees to each other.

Reynolds Numbers. A numerical ratio of the dynamic forces of mass flow to the shear stress due to viscosity.

RFCI. Resilient Floor Covering Institute.

RF Curing. Radio Frequency curing.

RFP. Request for Proposal.

RH. 1. Right hand. 2. Right Hand, a door handing designation.

Rheology. A science dealing with the deformation and flow of matter.

Rheostat. An electrical instrument used to control a current by varying the resistance.

RHMS. Round Head Machine Screw.

Rhombus. A polygon with four equal sides that do not meet at right angles.

RHR. Right Hand Reverse, a door handing designation.

RHWS. Round Head Wood Screw.

Rib. 1. A plain or molded arched member which forms a support for an arch or vault. 2. A decorative feature resembling a rib on the surface of a vault or ceiling; a projecting piece of molding upon the interior of a vault or used to form tracery in ornamental work.

Ribband. 1. The support for the second-floor joists of a balloon-frame house.2. The horizontal member of a wood frame wall which supports floor joists or roof rafters; also called a Ribbon.

Ribbed Shells. Spatial structures with material placed primarily along certain preferred rib lines, with the area between the ribs filled with thin slabs or left open.

Ribbon. See Ribband.

Rib Lath. Metal lath with ribs formed by folding to provide stiffness for greater span.

Richter Scale. A logarithmic scale that is used to measure the magnitude of earthquakes.

Riddle. A coarse sieve, as for sand or gravel.

Ride the Brush. To bear down on the brush to the extent that the paint is applied with the sides of the bristles instead of the flat ends; this shortens the life of the brush.

Ridge. 1. The top edge or corner formed by the intersection of two roof surfaces. 2. The board against which the tips of rafters are fastened.

Ridge Board. The board against which the tips of rafters are fastened; the top line of a roof; the ridge.

Ridge Cut. Any cut made in a vertical plane; the vertical cut at the top end of a rafter.

Ridge Rib. A rib forming the ridge in a vaulted roof.

Ridge, Roof. See Roof Ridge.

Ridge Vent. A vent mounted along the ridge of a roof to aid in ventilating an attic space.

Ridging. 1. An upward, tenting displacement of a roofing membrane, frequently occurring over insulation joints, deck joints, and base sheet edges. 2. See Beading.

Riffler. A small handtool for filing, scribing, or scraping.

Rifle Range. A place where rifle shooting is practiced.

Rift Valley. A graben.

Right Angle. An angle of 90 degrees.

Right Triangle. A triangle having one right angle.

Right-of-Way. The legal right, usually by easement, to pass over land owned by another.

Rigid Copper Tubing. Hard copper pipe used when installing water lines, particularly where they can be seen.

Rigid Frame. Two columns and a beam or beams attached with moment connections; a moment-resisting building frame; the connections between the beam and columns are constructed so as to transmit moments.

Rigidity. Relative stiffness of a structure or element.

Rigid Insulation. Thermal insulating material made of polystyrene, polyurethane, polyisocyanurate, cellular glass, or glass fiber, sometimes offered with a skin surfacing, formed to flat board shape of constant thickness.

Rim. A designation of articles of hardware designed for application to the surface of doors and windows, such as rim locks.

Ringelmann Sale. Device for measuring smoke density.

Ring Nail. See Annular Ring Nail.

Ring, Split. See Split Ring.

Ring, Toothed. See Toothed Ring.

Rip. To cut lumber lengthwise, parallel to the grain.

Riparian Rights. The rights of an owner whose land adjoins water, such as a river, stream, lake, or sea.

Rip Fence. A guide on a table saw to assist in cutting lumber in the direction parallel to the grain.

Ripper. Narrow strips of gypsum board used for soffits, window returns and finished openings.

Ripping. Sawing wood along the grain.

Ripple Effect. The effect that escalation or delay has on other work in a construction schedule; impact.

Riprap. A foundation or sustaining wall of stones placed together without order, as in deep water or on an embankment, to prevent erosion.

Ripsaw. A saw designed for cutting lumber length wise.

RIS. Redwood Inspection Service.

Rise. The vertical distance through which anything rises, as the rise of a roof or stair.

Riser. 1. A single vertical increment of a stair; the vertical face between two treads in a stair. 2. A vertical run of plumbing, wiring or ductwork. 3. A spacer between shipping units to allow entry of fork lift blades. 4. The flow channel or pipe that distribute the heat transfer liquid across the face of an absorber.

Riser Band. A flat, horizontal member usually ornamental, of relatively slight projection, mounted on a riser between two stair treads.

Riser Hanger. The horizontal member of a stair structure which supports the riser.

Riser Valve. Device used to manually control flow of refrigerant in vertical piping.

Risk. Chance or possibility of danger, loss, injury, or other adverse consequences.

Risk Management. The taking of precautions to avoid unexpected property and casualty losses which can severely damage the business's prospects; insurance is one of the best means of reducing the company's exposure to risk.

Rivet. A structural fastener on which a second head is formed after the fastener is in place.

Rivet Pattern. The arrangement, spacings, and dimensions of rivets used to attach two or more structural members together.

RMA. Rubber Manufacturers Association.

Ro. Rough.

Road Oil. A heavy petroleum oil, usually one of the Slow-Curing (SC) grades.

Roadway Excavation. The act or process of removing or moving obstructions such as boulders, rocks, earth to prepare a path for a road or highway.

Robot. A mechanical device that is operated by automatic controls and is used in manufacturing processes to perform functions previously carried out by human workers.

Robotics. The science concerned with the design, manufacture, and utilization of robots.

Rock. The hard, firm, and stable parts of earth's crust; any material which requires blasting before it can be dug by available equipment.

Rock Anchor. A post-tensioned rod or cable inserted into a rock formation for the purpose of anchoring a structure to it.

Rock Asphalt. See Asphalt Rock.

Rock Drilling. The act or process of boring holes into rock.

Rock Flour. Crushed rock in fine particles like powder.

Rock Gun. A device for throwing aggregate onto a soft bedding coat in applying marblecrete.

Rock Pocket. A concrete defect in which a group of aggregate is not adequately surrounded by cement paste

Rockwool Insulation. Insulation made by forming fibers from molten rock and slag; used as insulation in ceilings and walls.

Rococo. An artistic style of the 18th century that is characterized by fanciful, excessive, intricate, and elaborate ornamentation; baroque.

Rod. 1. A slender stick, pole, or bar of wood or metal. 2. A unit of length, used in land measuring, of 16.5 feet. 3. A square rod, 273.9 square feet.

Rodding Off. See Floating.

Rod, Ground. See Ground Rod.

Rod, Guy. See Guy Rod.

Rod, Hanger. See Hanger Rod.

Rod Saw. A saw consisting of a steel rod approximately 1/8" in diameter used to cut circles or irregular curves in tile; the rod has tungsten carbide particles embedded in the surface.

Rod, Tamping. See Tamping Rod.

Rod, Traverse. See Traverse Rod.

Rolled. A brick that is laid with an overhanging face.

Rolled Steel Section. Any hot-rolled steel member such as a wide flange beam, I-beam, angle, channel, tee, or similar sections.

Roller Blind. A window covering that retracts to the top of the window by means of a spring-loaded roller.

Roller Coating. 1. A paint or other material applied to the surface by a paint roller. 2. Process of finishing an article by means of hard rubber or steel rollers.

Roller Latch. A friction door latch employing a roller under spring tension which engages a recessed strike plate formed to receive the roller.

Roller Marks. Surface defects caused during the manufacturing process as the gypsum board is moved through the various production operations.

Roller Shade. A window shade on spring-loaded a roller that allows the shade to go up and down.

Roller Strike. A door latch strike having a rolling member at the point of latch bolt contact to minimize friction.

Roll Roof. A roof that has been covered with an asphaltic material that comes in rolls.

Roll Roofing. Asphaltic roofing, coated felts, either smooth or mineral surfaced, that comes in rolls;

Roll-Up Door. A door which raises and rolls into a coiled configuration and lowers on tracks on either jamb.

Rolok Wall. An economy wall and a substitute for a solid brick wall, it is a combination of several shells of brickwork laid flat or on edge to produce a cellular type of open construction.

ROM. Read Only Memory.

Roman Arch. A round topped arch.

Roman Brick. Brick whose nominal dimensions are 3-1/2 x 2 x 12 inches.

Roman Cities. Influenced by the plan used for military camps, grid iron city blocks; more formal organization of public spaces in the nucleus of city.

Romanesque. A style of architecture prevalent in Europe in the 9-12th century, with massive vaulting, round arches, decorative use of arcades, and profuse ornament.

Roman Numeral. 1. Any of the Roman letters representing numbers, I = 1, V = 5, X = 10, L = 50, D = 500, M = 1000. 2.

A numeral in the system of notation that is based on the ancient Roman system.

Rood Screen. A decorative screen separating the nave from the chancel.

Roof. 1. The top cover of a building or structure. 2. The top limit on price, a Ceiling Price.

Roof Adhesive. A bonding agent used to cement roof materials.

Roof, Asphalt. See Asphalt Roof.

Roof Assembly. An assembly of interacting roof components, including the roof deck, designed to weatherproof and, normally, to insulate a building's top surface.

Roof Ballast. Crushed rock or gravel which is spread on a roof surface to form its final surface.

Roof Bond. A legal guarantee that a roof installed is in accordance with specifications and will be repaired or replaced if it fails within a certain period of time due to normal weathering.

Roof, Built-Up. See Built-Up Roof

Roof Cricket. Relatively small elevated area of wood roof constructed to divert water around chimney, curb, or other projection.

Roof, Curb. See Gambrel Roof.

Roof Decking. 1. Two inch or thicker lumber installed on a roof system. 2. The structural skin of a roof.

Roof Drain. 1. A drainage outlet on a roof. 2. A drainage conduit or pipe that collects water runoff from a roof and leads it away from the structure.

Roofer. A person who constructs or repairs roofs.

Roof Expansion Joint. A break or space between roofing materials to allow for thermal expansion and contraction.

Roof Flashing. A weather shield commonly used at penetrations of a roof by pipes or columns to insure no passage of the elements.

Roof, Fluid-Applied. See Fluid-Applied Roof.

Roof Framing. The skeleton structure of a roof system.

Roof Hatch. Fabricated, usually of metal, horizontal access door mounted on curb to permit access to flat or low sloped roofs from the building interior.

Roof, Hip. See Hip Roof.

Roofing Cement. A general term for a variety of trowelable mastics, asphalt, or tar that are used in roof construction and repair.

Roofing. 1. Materials for a roof. 2. The process of constructing a roof. 3. The skilled trade of applying roofing; a Roofer.

Roofing Consultant. An expert in the design, specification, construction, and repair of roofs.

Roofing Felt. A felt sheet impregnated with a bituminous waterproofing material.

Roofing Nail. A short, often galvanized, nail with a broad head.

Roofing Removal. The act or process of stripping off an old roof system.

Roofing Square. 1. An area of 10 by 10 feet of roofing surface. 2. 100 square feet of roofing surface.

Roofing System. A group of interacting roof parts designed to weatherproof and to insulate a roof.

Roof Insulation. 1. Materials used between rafters or roof supports for the protection from heat and cold. 2. Solid sheets of insulating material installed on a flat roof.

Roof Panel. Roof panel construction from boards, plywood, or metal sheets that can be installed as a pre-assembled unit.

Roof Ridge. The uppermost peak of a roof where the roof planes meet.

Roof Scupper. See Scupper, 1.

Roof Sheathing. The first layer of covering on a roof, fastened to the rafter boards, used to support the roofing material.

Roof Shingle. Pieces of wedge shaped wood or other material used in overlapping courses to cover a roof, usually made of wood, asphalt, clay tile, or metal.

Roof Slab. The flat section of a reinforced concrete roof, supported by beams, columns, or other framework.

Roof Specialties. A fitting or piece of trim used in the installation of a roof, such as gravel stop, flashing, or vent strips.

Roof Spinner. An exhaust ventilation apparatus mounted on a roof which uses the wind to create an upward draft and thus create ventilation.

Roof System. A general term for the complete assembly of waterproof covering, insulation, vapor barrier, ballast, and substrate.

Rooftop AC. Air-conditioning system or equipment installed on a roof.

Rooftop Heater. A heating unit installed or placed on a roof.

Rooftop Unit. A system of conditioning air in a self contained unit, installed directly above the conditioned space, in a weatherproof enclosure.

Roof Vent. 1. A usually screened opening on a gable end or roof structure to allow a free flow of ventilation air into the underroof area. 2. The unit mounted on a roof that is the terminus of a ventilation pipe.

Roof Walkway. A permanent aisle for safe access across a roof, also serves as a protection for the roofing material when maintenance is being done.

Rooming House. A hotel.

Room Resonance. The sound resonance of an enclosed space, which depends in the relationship between the sound wavelength, the space dimensions, and the reflective surfaces.

Root Ball. The soil and earth that clings to roots of a dug up tree that is later to be planted in a different location.

Root Buttress. A root that is above ground where it joins the trunk of a tree.

Root Face. In welding, that portion of the groove face adjacent to the root of the joint.

Root Hook. A very heavy hook designed to catch and tear out big roots when it is dragged along the ground.

Root of Joint. That portion of a joint to be welded where the members are closest to each other.

Root of Weld. The point at which the bottom of the weld intersects the base-metal surfaces and is the farthest from application of weld heat and/or filler metal side.

Root Opening. In welding, the separation between the members to be joined at the root of the joint.

Rope. A stout cord made by twisting or braiding together strands of hemp, sisal, flax, cotton, nylon, wire, or similar material.

Rose. A trim plate attached to the door hardware behind the knob.

Rosette. 1. A circular decorative ornament made of wood, metal, or plaster. 2. A metal trim at a ceiling mounted lighting fixture or at a blanked-off lighting fixture outlet.

Rose-Window. A round window filled with tracery and, usually, stained glass; characteristic of Gothic churches.

Rosewood Veneer. A thin layer of cabinet wood with a dark red or purplish color streaked and variegated with black, applied to give a superior and decorative surface.

Rosin. A hard resin obtained from pine trees containing principally isomers of abietic acid; wood rosin is obtained from stump or dead wood, using steam distillation; gum rosin is obtained from the living treerosin is used in making varnish, soap, and solder flux.

Rot. Decay.

Rotary Blade Compressor. Mechanism for pumping fluid by revolving blades inside a cylindrical housing.

Rotary Compressor. See Compressor, Rotary.

Rotary-Cut Veneer. Veneer cut in a lathe which rotates a log or bolt, chucked in the center, against a knife.

Rotate. To turn or revolve about an axis or center.

Rotation. In building structures, the torsional movement of a diaphragm about a vertical axis.

Rotisserie. 1. A restaurant specializing in roasted and barbecued meats. 2. A cooking device with a revolving spit for roasting and barbecuing meat.

Rotor. A rotary part of a machine.

Rottenstone. A siliceous limestone which when finely pulverized, is used in wood finishing; it has negligible cutting action but is fine for polishing; also called Tripoli.

Rough-In. All work that must be performed before plastering.

Rough Horse. See Carriage.

Rough Lumber. Lumber that has not been dressed (surfaced) but which has been sawed, edged , and trimmed.

Rough Opening. The clear dimensions of the framed opening that must be provided in a wall to accept a given door, window unit, or other unit.

Round. 1. Circular. 2. A wooden dowel.

Round Edge. A preformed factory rounded edge design.

Round Head Wood Screw (RHWS). A wood screw with a rounded head protruding above the surface; driven with a screw driver; usually installed with a washer under the head.

Round Pattern. Circular spray pattern.

Round Wire. See Loop Pile.

Round, Quarter. See Quarter Round.

Rout. To gouge out or make a furrow in, as in wood.

Router. Woodworking tool with a high speed bit for cutting rebates, dadoes, insetting butts, and forming decorative or rounded edges.

Row House. One of a series of houses connected by common sidewalls and forming a continuous group.

Rowlock Course. A course of headers laid on their edges instead of flat side; rows of brick laid on edge especially used for the ring of an arch.

Rows. The number of lengthwise yarn tufts in one inch of Axminster or Chenille carpet; compare with Wires.

RPM. Revolutions Per Minute.

RTU. Roof Top Unit.

Rub Joint. A glue joint made by carefully fitting the edges together, spreading glue between them, and rubbing the pieces back and forth until the pieces are well rubbed together.

Rubber. A tough elastic material made from the latex of rubber trees, a natural polymer of a hydrocarbon.

Rubber Emulsion Paint. Paint with a vehicle of fine droplets of natural or synthetic rubber dispersed in water.

Rubber, Laminated. See Laminated Rubber.

Rubber Nosing. The projection of a rubberized stair tread over a riser in a stair system.

Rubber Set. See False Set.

Rubber Spacers. Cross and tee-shaped objects used to space tile on floors or walls; manufactured in thicknesses of 1/16", 1/8", 1/4", 3/8", and 1/2".

Rubber Trowel. A non-porous synthetic-rubber-faced float mounted on an aluminum back with a wood handle, used in tile grouting to force material deep into tile joints and to remove excess grout.

Rubbing Compound. An abrasive material used to produce a smooth finished wood surface.

Rubbing Oil. Neutral, medium-heavy mineral oil used as a lubricant for pumice stone in rubbing varnish and lacquer.

Rubbing Stone. A carborundum stone that is used to smooth the rough edges of tile.

Rubbing Varnish. A hard-drying varnish which may be rubbed with an abrasive and water or oil to a uniform leveled surface.

Rubbish Handling. The act or process of transporting or removing waste materials.

Rubbish Removal. The process of removing waste or garbage from an area.

Rubble. 1. Roughly broken quarry stone. 2. Rough broken stones or bricks used to fill in courses of walls or for other filling;

Rubble Masonry. Uncut stone, used for rough work, foundations, backfilling, and the like.

Rug. A term used to designate soft floor coverings laid on the floor but not fastened to it; usually a rug does not cover the entire floor.

Rule-of-Thumb. A rule for general guidance based on experience rather than on precision or scientific theory.

Ruler. A straight, usually graduated, piece of wood, plastic, or metal, used for drawing straight lines or for measuring.

Run. 1. In a stairway, the width of a step, measured from the face of one riser to the face of the next, and not including the nosing. 2. The horizontal distance covered by a flight of steps. 3. The length of the horizontal projection of a sloping member such as a rafter when in position. 4. A temporary planked scaffold for walking or wheeling materials. 5. That portion of a pipe or fitting continuing in a straight line in the direction of flow in the pipe to which it is connected. 6. An appreciable length of straight or nearly straight piping.

Rung. A step on a ladder.

Runner. Metal or wood track or strips placed at floor and ceiling to receive framing members, such as metal or wood studs.

Runner Channel. A steel channel member from which furring channels and lath are supported in a suspended plaster ceiling.

Runners. See Main Runners.

Running Bond. Brickwork consisting entirely of stretchers, lapping of units in successive courses so that the vertical head joints lap.

Running Mold. Constructed by plasterers in place by use of a metal pattern to form a molding of fresh plaster.

Running Time. Amount of time a condensing unit is run per hour or per 24 hours.

Running Track. An oval-shaped track for use by athletes in various running events.

Running Trap. A plumbing trap in which the inlet and outlet are in a straight horizontal line between which the water way is depressed to below the bottom of either the inlet or outlet.

Running Winding. Electrical winding of motor which has current flowing through it during normal operation of motor.

Runoff. The excess of rainwater or snowmelt that is not absorbed into the earth and drains to streams, rivers, or storm water collection devices.

Runs. In painting, curtains, sags, and other irregularities due to uneven flow.

Rural Letter Box. A roadside receptacle usually found in rural areas for delivery of mail.

Rust. A reddish or yellowish-brown coating formed on iron or steel caused by oxidation, usually as a result of moisture; ferrous oxide.

Rusticated Concrete. Beveled edges of concrete making the joints conspicuous.

Rustification. In building and masonry, the use of squared or hewn stone blocks with roughened surfaces and edges deeply beveled or grooved to make the joints conspicuous.

Rust-Inhibitive Washes. Solutions which etch the metal and form a dull gray coating of uniformly fine texture, thus producing rust-inhibitive surface receptive to priming coat.

Rutile. A mineral form of titanium oxide, TiO_2, tetragonal crystallization, but usually produced chemically for use in ceramics and other products.

R-Value. The number of minutes (seconds) required for 1 Btu (joule) to penetrate one square foot (square meter) of a material for each degree of temperature difference between the two sides of the material. The resistance of a material to the passage of heat. The reciprocal of conduction (1/c).

RW. Redwood.

Rwd. Redwood.

S

S&P. Shelf and pole.

S1E. Lumber, surfaced one edge.

S1S. Lumber, surfaced one side.

S1S1E. Lumber, surfaced one side and one edge.

S1S2E. Lumber, surfaced one side and 2 edges.

S2E. Lumber, surfaced two edges.

S2S. Lumber, surfaced two sides.

S4S. Lumber, surfaced 4 sides.

SA. Supply Air.

Sabin. The unit of acoustic absorption; one sabin is the absorption of one square foot of perfect sound-absorbing material.

Sack. A quantity of cement, 94 pounds, I cubic foot, in the United States for portland or air entraining portland cement or as indicated on the sack for other kinds of cement.

Sacking. Removing or alleviating defects on a concrete surface by applying a mixture of sand and cement to the moistened surface and rubbing with a coarse material such as burlap.

Sack Joint. Joint that has been wiped or rubbed with a rag or object such as a rubber heel.

Saddle. 1. A built up section of the roof substrate to divert water toward the drains. 2. See Threshold, 2.

Saddle Board. The finish of the ridge of a pitched roof house; also called Comb Board.

Saddle Fitting. A fitting used to install a branch line from an existing run of pipe; first a hole is made in the pipe, then the saddle is clamped around the hole.

Saddle Tie. A specific method of wrapping hanger wire around main runners or of wrapping tie wire around the juncture of main runner and cross furring.

Saddle Valve. Valve body shaped so it may be silver brazed or clamped on to a refrigerant tubing surface; also called Tap-a-Line.

Safeflower Oil. Oil from seed of thistle-like plant grown mostly in Egypt and India.

Safe, Office. See Office Safe.

Safety Arch. See Discharging Arch.

Safety Belt. A solid belt that may be attached to a secured lifeline.

Safety Can. Approved container of not more than 5-gallon capacity, with a spring-closing lid and spout cover; designed to relieve internal pressure safely when exposed to fire.

Safety Chain. Chain installed horizontally in railing assembly to provide for ease in providing temporary opening in railing.

Safety Control. Device which will stop the refrigerating unit if unsafe pressures, temperatures, or dangerous conditions are reached.

Safety Factor. See Factor of Safety.

Safety Glass. Specific type of glass having the ability to withstand breaking into large jagged pieces, usually tempered and laminated.

Safety Glasses. Plastic glasses that give protection to the wearer doing work that might send off particles of matter, such as when grinding and chipping.

Safety Motor Control. Electrical device used to open circuit to motor if temperature, pressure, or current flow exceed safe conditions.

Safety Net. A woven meshed fabric that is suspended below a construction activity to protect materials and people that may fall from dangerous heights.

Safety Nosing. Stair nosing with abrasive non-slip strip surface flush with tread surface.

Safety Plug. Device which will release the contents of a container before rupture pressures are reached.

Safety Relief Valve. A valve designed to release water or steam if the temperature and pressure have risen to the point that an explosion could occur.

Safety Switch. A mechanism on a machine or device that prevents that machine from being operated until the mechanism is deactivated.

Safety Tread. Stair tread which is covered on the top surface with abrasive or non-slip material.

Safety Valve. Pressure release valve preset to be released when pressure exceeds safe operating limit.

Safing. Fire-resistant material inserted into a space between a curtain wall and a spandrel beam or column, to retard the passage of fire through the space.

Sag. 1. A fresh plaster wall surface has developed a slide. 2. An unevenness or irregularity in a coat of paint, varnish, or lacquer, resulting from too much of the liquid collecting in one spot or area.

Sag Pond. A pond occupying depression along a geological fault; the depression is normally due to uneven settling of the ground or other causes.

Salamander. A portable source of heat in a building under construction, customarily kerosene or oil-burning, used to temporarily heat an enclosure; often used around newly placed concrete to prevent freezing.

Sale-Leaseback. A financing arrangement in which an owner sells real property to an investor and leases it back, usually for the purpose of freeing invested capital.

Sales Tax. A tax levied on sales of goods and services which is calculated as a percentage of the purchase price and collected by the seller.

Salient Angle. Projecting outward from a line, surface or level; the opposite of a reentrant angle.

Saline. A solution of salt in water.

Saline Water. Those waters having a specific conductivity in excess of a solution containing 6,000 parts per million of sodium chloride.

Sally Port. A gate or passage in a fortified place for use by troops making a sortie.

Salmon Brick. Relatively soft, under-burned brick, so named because of its color; used mostly where it is protected from the weather and where it does not have to support a great weight.

Salt. 1. Sodium chloride, NaCl, used for preserving the freshness of food. 2. Substance that results from reaction between acid and base.

Salt Glaze. A glaze produced by the reaction, at elevated temperature, between the ceramic body surface and salt fumes produced in the kiln atmosphere.

Saltpetering. See Efflorescence.

Salt Spray. A salt fog test environment.

Salvage. The saving and utilization of waste materials leftover from a remodeling job or a fire.

Salvage Value. The estimated value of an asset at the end of its economic life.

Samples. Material samples requested by the architect of the general contractor or materials specified for the project.

Sample Soil. A representative specimen of soil from a site.

Sampling. The method of obtaining small amounts of material for testing from an agreed-upon lot.

San Blast. A system of abrading a surface such as concrete by a stream of sand, or other abrasive, ejected from a nozzle at high speed by water and/or compressed air.

Sand. Loose granular material resulting from the natural disintegration of rock or from the crushing of friable sandstone, passing through a #4 sieve but predominantly retained on a #200 sieve; manufactured sand is the fine material resulting from the crushing and classification by screening, or otherwise, of rock, gravel or blast furnace slag.

Sand Asphalt. A mixture of sand and asphalt cement or cutback or emulsified asphalt; it may be prepared with or without special control of aggregate grading and may or may not contain mineral filler; either mixed-in-place or plant mix construction may be employed; sand asphalt is used in construction of both base and surface courses.

Sandbag. A bag filled with soil or sand to act as a temporary barricade or dam.

Sandblast. A system of cutting or abrading a surface such as concrete by a stream of sand ejected from a nozzle at high speed; compressed air is used to propel a stream of wet or dry sand onto the surface; often used for cleanup of horizontal construction joints or for exposure of aggregate in architectural concrete; a method of scarifying the surface of concrete or masonry to provide a bondable surface; used to clean metal before painting.

Sandblast Masonry. The act or process of abrading or cutting masonry structure surfaces, using sand ejected from a nozzle at high speed by compressed air; effective form of cleaning masonry, but used sparingly because of its inclination to remove the outer, weather resistant, usually glazed or polished, surface.

Sand Cushion Terrazzo. Terrazzo with an underbed that is separated from the structural floor deck by a layer of sand.

Sand Down. Remove the gloss of an old surface finish and smooth it prior to refinishing.

Sander. A machine that smoothes or polishes by means of abrasive material usually in the form of a disk or belt; Disk sander or Belt Sander.

Sand Filter. Pool filter using sand as filtering medium.

Sand Finish. Rough finish plaster wall.

Sand Holes. Tiny pits in the surface of ceramic tile.

Sanding. Rubbing sandpaper or similar abrasive over a surface before applying a finish.

Sanding Pole. In gypsumboard joint finishing, a sandpaper holder affixed to the end of a handle with a swivel to aid in the sanding process.

Sanding Sealer. A lacquer used as a seal coat over a filler; usually given some filling action by adding inert substances.

Sand Mold Brick. See Soft Mud Brick.

Sandpaper. Paper with sand or other abrasive stuck to it for smoothing or polishing.

Sandstone. A sedimentary rock formed from sand.

Sandwich Glass. A pane of glass that contains an inner layer of material between two outer layers of glass to provide additional insulating properties or resistance to breakage.

Sandwiching. In manufacturing gypsumboard, forming the gypsum core between two plys of paper.

Sandwich Panel. A panel consisting of two outer faces of wood or metal, bonded to a core of insulating foam.

Sandy Finish. A surface condition having the appearance of sandpaper; may result from overspray; or a finish with sand or walnut shells.

Sanitary Facility. Any single unit or a combination of water closets, urinals, lavatories, bathtubs or showers, together with the room or space in which they are housed.

Sanitary Piping. Drain, waste, and vent plumbing systems.

Sanitary Sewer. A sewer used only for carrying the liquid or water-borne wastes from plumbing fixtures; does not include storm, surface, or ground water.

Sanitary Tee. A fitting in a DWV plumbing system.

Sanitizer. One of three groups of antimicrobials registered by EPA for public health uses; EPA considers an antimicrobial to be a sanitizer when it reduces but does not necessarily eliminate all the microorganisms on a treated surface; to be a registered sanitizer, the test results for a product must show a reduction of at least 99.9% in the number of each test microorganism over the parallel control.

Sap. Most of the fluids in a tree; certain secretions and excretions, such as oleoresin, are excepted.

Saponification Number. Number of milligrams of potassium hydroxide needed to neutralize the acid in one gram of substance after it has been saponified.

Saponify. To convert an oil or a fat into soap by action of an alkali; when linseed oil paint comes in contact with a surface that contains strong alkali and water, like new concrete floor, the oil is saponified and loses its bonding qualities.

Sapwood. The wood of pale color near the outside of the log; under most conditions the sapwood is more susceptible to decay than heartwood.

Sarcophagus. A stone coffin.

Sash. The framework which holds the glass in a window.

Sash Balance. A spring-loaded device for counterbalancing the weight of a vertical sliding sash as on a double-hung window.

Sash Bead. A strip with one edge molded, against which a sash slides.

Sash Chain. A metal chain adapted for use with vertical sliding sash, attached to the sash and to the counterbalancing sash weight.

Sash Cord. Cord or rope used similarly to sash chain.

Sash Pole. A wood or metal pole to which a sash pole hook is attached, used to raise or lower a transom or sash beyond hand reach.

Sash Pulley. A pulley mortised into a double-hung sash frame over which the sash cord or sash chain passes.

Sash Socket. A metal plate, attached to a sash, containing a hole or cup to receive a sash pole hook.

Sash Tool. A beveled paint brush for use on windows and other precise cutting in.

Sash, Transom. See Transom Sash.

Sash Weight. A cast iron weight used for counterbalancing the weight of a vertical sliding sash as on a double-hung window.

Satin Finish. The dried film of paint or other finishing material which does not have a full luster, but a dull luster like that of satin.

Saturated Air. Air that contains all of the moisture vapor it can hold; in saturated air, the dry-bulb, wet-bulb, and dew point temperatures are all equal.

Saturated Felt. A felt that is partially impregnated with low softening point bitumen.

Saturated Surface Dry (SSD). A condition of an aggregate which holds as much water as it can without having any water within the pores between the aggregate particles.

Saturated Vapor. Vapor condition which will result in condensation into droplets of liquid if vapor temperature is reduced.

Saturation. Condition existing when substance contains all of another substance it can hold for that temperature and pressure.

Saturation Coefficient. See C/B Ratio.

Savanna. A grassy plain in tropical and subtropical regions, with few or no trees.

Saw, Band. See Band Saw.

Saw Cut, Concrete. See Concrete Saw Cut.

Saw Cutting, Concrete. See Concrete Saw Cutting.

Sawed Control Joints. A joint cut in hardened concrete, generally not to the full depth of the member, by means of special equipment.

Sawed Veneer. Veneer produced by sawing.

Saw Fence. A device on a table saw for guiding lumber being sawed.

Sawhorse. A frame, usually four-legged, for supporting wood while sawing.

Saw Kerf. 1. A groove in a piece of wood made by a saw blade. 2. That portion of a log, timber, or other piece of wood removed by the saw in parting the material into two pieces.

Saw-Tooth Roof. A roof shape that is saw-toothed in section; the vertical planes are usually to accommodate windows.

SBA. Small Business Administration.

SBCCI. Southern Building Code Congress International, Inc.

SBS. Sick Building Syndrome.

Scab. 1. A short piece of lumber used to splice two other pieces together. 2. One who refuses to join a labor union or who participate in practices in opposition to the union.

Scaffold. 1. A temporary structure or platform for workmen to stand or sit on when working in high places; also called Staging. 2. A temporary framework structure for the support of concrete forms and other shoring; adjustable metal scaffold is frequently adapted for shoring of concrete formwork.

Scaffolding. The materials used in constructing scaffolds.

Scagliola. An imitation marble made by the plasterer; composed of a combination of Keene's cement, glue, isinglass, and coloring material; it takes a high durable polish.

Scale. 1. A proportion between two sets of dimensions, as between those of a drawing and its original; for example, the scale of a drawing may be expressed as 1/4 inch = one foot. 2. A measuring tool used by architects and engineers in preparing drawings to a proportionate scale. 3. To measure a drawing with a scale. 4. Either pan or tray of a balance. 5. To climb, as a ladder. 6. A series of graduated marked spaces for measuring something, as on a thermometer. 7. Rust occurring in thin layers. 8. Hard deposit of minerals on heater coils and pool surfaces.

Scale Drawing. A drawing in which the measurements represented are drawn to a predetermined scale, such as 1/4 inch equals one foot, so that all elements and dimensions in the drawing are proportional in length and width to the actual room, floor, or building depicted.

Scalene Triangle. A triangle having three unequal sides.

Scaler. A hand cleaning chisel.

Scaling. 1. Finish condition in which pieces of the dried finishing material come off exposing the surface below. 2. Prying or chipping loose pieces of rock off a building face or roof to avoid danger of their falling unexpectedly. 3. The breaking away of a hardened concrete surface, usually to a depth of 1/16 to 3/16. 4. Measuring a drawing with a scale.

Scallops. The up-and-down uneven effect along the edge of carpet caused by indentations where tacks are driven.

Scantling. 1. A small piece of lumber. 2. Lumber with a cross-section ranging from 2 by 4 inches to 4 by 4 inches.

Scarf Joint. 1. A glued end connection between two pieces of wood, using a sloping cut to create a larger surface for the glue bond, to allow it to develop the full tensile strength of the wood it connects. 2. A timber spliced by cutting various shapes of shoulders, or jogs, which fit each other. 3. A joint between two pieces of wood which allows them to be spliced lengthwise.

Scarfing. A joint between two pieces of wood which allows them to be spliced lengthwise.

Scarifier. A piece of thin metal with teeth or serrations cut in the edge, used to roughen fresh mortar surfaces to achieve a good bond for the tile; a scarifier also can be used to roughen the surface of concrete.

Scarlet Lake. Pigment made by precipitation of aniline color upon base of alumina hydrate and barium sulphate.

Scarp. 1. A cliff, escarpment, or steep slope of some extent formed by a geological fault. 2. A cliff or steep slope along the margin of a plateau, mesa, or terrace.

Scarred Faces. Surface blemishes caused by scraping or other marring of ceramic tile.

SC Asphalt. See Slow-Curing Asphalt.

Scavenger Pump. Mechanism used to remove fluid from sump or container.

SC Grade. See Road Oil.

Schedule. 1. A list, catalog, or inventory. 2. A time table of things to be done. 3. A pipe size system originated by the iron industry, giving outside diameters and wall thickness of iron and plastic pipes.

Scheduling Software. Computer software used to prepare a construction schedule.

Schematic Design Phase. One of the standard phases of architectural service (Schematic Design Phase, Design Development Phase, Construction Documents Phase, Bidding or Negotiation Phase, and Construction Phase-Administration of the Construction Contract).

Schematic Diagram. A drawing showing general layout and relationships of the elements, not necessarily to scale.

Schematic Drawings. Drawings prepared by architects for the Schematic Design Phase.

Schrader Valve. Spring-loaded device which permits fluid flow in one direction when a center pin is depressed; in other direction when a pressure difference exists.

Scissors Lift. A low-rise lift utilizing a scissorlike mechanism to raise the platform.

SCL. Structural Composite Lumber.

Sconce. A wall bracket candle holder or lighting fixture.

Score. To cut a surface with a sharp blade before breaking.

Scoreboard, Gym. See Gym Scoreboard.

Scotch. See Scutch.

Scotch Yoke. Mechanism used to change reciprocating motion into rotary motion or vice-versa; used to connect crank-shaft to piston in refrigeration compressor.

Scotia. A concave molding.

SCPI. Structural Clay Products Institute.

SCR. 1. Silicon-Controlled Rectifier. 2. Structural Clay Research, trademark of the Structural Clay Products Research Foundation.

Scratch. A defect in a surface made by scraping with a sharp pointed object.

Scratch Coat. The first coat of plaster applied to a surface in three coat work; the term originates from the practice of cross-raking or scratching the surface of this coat with a comb-like tool to provide a mechanical key to aid bond with the brown coat.

Scratcher. Any serrated or sharply tined object that is used to roughen the surface of one coat of mortar to provide a mechanical key or bond for the next coat.

Scratching. The application of a scratch coat and its combing with a scratcher.

Screed. 1. A strip of wood, metal, or plaster that establishes the level to which concrete or plaster will be placed. 2. To strike or plane off wet mortar or concrete which is above the desired plane or shape.

Screed Guide. Firmly established grade strips or side forms for unformed concrete which will guide the strikeoff in producing the desired plane or shape.

Screen. 1.A fixed or movable partition for separating, concealing, or sheltering from drafts, excessive heat or light, or from observation. 2. A white or silver surface for the projection of films. 3. A wood or metal frame with fine wire netting to keep out insects. 4. See Sieve.

Screen Door Latch. A small locking or latching device used on screen doors and operated by a knob or a lever handle.

Screen Door. A door consisting of metal or plastic insect screen stretched taut on a light wood or metal frame.

Screw. A spirally grooved metal fastener installed with a screwdriver.

Screwable Type Stud. Metal stud fabricated from not lighter than 26 gauge metal with knurled flanges to facilitate easier penetration of self-tapping screws or divergent point staples.

Screw Conveyor. A helical conveyor.

Screw Cover, Box. See Box, Screw Cover.

Screwdriver. A hand tool for installing screws.

Screw Gun. A hand held device resembling an electrical drill designed or adapted to mechanically drive and set screws.

Screw Jack. A lifting mechanism consisting of a threaded shaft for raising heavy loads.

Screwless Rose. A door knob rose that has a concealed method of attachment.

Screw Pump. Compressor constructed of two mated revolving screws.

Screw Ring. A metal or plastic ring which is threaded on its inside and used as a connector.

Screw, Self-Tapping, Drywall. See Self-Tapping Drywall Screw.

Scribe. To score or mark along a cutting line.

Scribing. 1. Transferring the exact irregularities of a wall or other surface onto a piece of carpet, wood or paper, which is then cut to fit those irregularities. 2. Fitting woodwork to an irregular surface; in moldings, cutting the end of one piece to fit the molded face of the other at an interior angle to replace a miter joint; such a joint is called a Coped Joint.

Scrim. 1. A fabric reinforcing that is used between moppings to strengthen parts of the roofing membrane. 2. A durable plain-woven fabric used for carpet backing. 3. Rough textured woven cloth worked into brown or finish coat to add crack resistance to the plaster; its use is now very rare.

Scrim Back. See Double Back.

Scroll. A convoluted or spiral ornamental design in metal or wood work.

Scrollsaw. A thin-bladed handsaw in a frame to keep the blade in tension, for cutting curves and irregular designs; a fretsaw

Scrollwork. A decoration consisting of spiral lines.

Scuffed Paper. A surface scrape usually caused by sliding one gypsum board over the face of another, or by sanding beyond the finished joint.

Sculptor. An artist who makes sculptures.

Sculpture. The art of creating forms, often representational, in the round or in relief by chiseling stone, carving wood, or casting metal.

Sculptured Carpet. A pattern formed when certain tufts are eliminated or pile yarns drawn tightly to the back to form a specific design in the face of the carpet; the result simulates the effect of hand carving; also called Carved Carpet.

Sculptured Tile. Tile with a decorative design of high and low areas molded into the finished face.

Scum Channel. Perimeter overflow inset into pool wall as surface water collection for filtration

Scupper. 1. An opening through a parapet wall through which water can drain from the edge of a flat roof. 2. An opening through an exterior wall through which water can drain.

Scutch. In masonry, a bricklayer's cutting tool used for dressing and trimming brick to a special shape; it resembles a small pick; also called a Scotch.

Scuttle. A small access opening in a wall, ceiling, or roof fitted with a lid.

SDI. 1. Steel Deck Institute. 2. Steel Door Institute

SE. Structural Engineer.

Seal. 1. Something that confirms, ratifies, or makes secure. 2. An amount of water standing in a drain trap to prevent foul air from rising; see Trap Seal. 3. Close securely or hermetically. 4. Apply a non-porous coating to a surface to make it impervious. 5. A counterflashing closure strip, narrow in shape, made of bituminous materials. 5. To secure a roof from the entry of moisture.

Sealable Equipment. Electrical equipment enclosed in a case or cabinet that is provided with means of sealing or locking so that live parts cannot be made accessible without opening the enclosure.

Seal Coating. Coating used to prevent excessive absorption of the first coat of paint by the substrate; a primer.

Sealant. 1. A material used for sealing spaces between materials. 2. An elastomeric material that is used to fill and seal cracks and joints; at expansion joints, this material prevents the passage of moisture and allows horizontal and lateral movement. 3. A caulking material used to fill voids or seams for acoustical, air infiltration, or thermal purposes. 4. A mixture of polymers, fillers, and pigments used to fill and seal joints where moderate movement is expected; it cures to a resilient solid.

Sealed Refrigeration Compressor. See Hermetic Compressor.

Sealed Unit. See Hermetic System.

Sealer. A base coating of paint to seal and equalize suction differences and prevent absorption of subsequent coats.

Sealing Compound. 1. Material used to exclude water and solid foreign materials from openings. 2. A substance that prevents water and moisture from entering joints.

Seal Leak. Escape of oil or refrigerant at the junction where a shaft enters a housing.

Seal, Shaft. See Shaft Seal.

Seam. 1. A line, groove, or ridge formed at the meeting of two edges of a material. 2. A layer of rock, coal or ore. 3. Treated gypsum board joints. 4. A joint between sections of carpet; see Back Seam, Face Seam, Cross-Seam, Length Seam, and Side Seam.

Seamless Gutter. Rainwater gutter without seams; usually made on a truck containing a forming machine that processes long coils of sheet metal on the building site.

Seam Weld. A weld consisting of a series of overlapping spot welds made by seam welding or spot welding.

Seasoned Loan. A loan that has had a good payment record over an extended period of time

Seasoned Lumber. Air- or kiln-dried undefective lumber that is properly stored and is ready for use.

Seasoning. Removing moisture from green wood to improve its serviceability.

Seat. 1. A piece of furniture for sitting on, such as a chair, stool, bench, or pew. 2. The portion of a chair where one sits. 3. That portion of a valve mechanism against which the valve presses to effect shutoff. 4. A special chair of one in authority.

Seat Cut. The cut at the bottom end of a rafter to allow it to fit upon the plate.

Seated Connection. A connection in which a steel beam rests on top of a steel angle fastened to a column or girder.

Seat of a Rafter. The horizontal cut on the bottom end of a rafter which rests upon the top of the plate.

Seat, Grandstand. See Grandstand Seat.

Second. 1. A unit of measurement of angles, equal to 1/60 of a minute. 2. A unit of measurement of time, equal to 1/60 of a minute. 3. A Second Mortgage. 4. A Second Trust Deed.

Secondary Branch. The second or more drain branch of the building drain.

Secondary Clay. See Sedimentary Clay.

Secondary Color. The result of mixing two primary colors; red and yellow yields orange; yellow and blue yields green; blue and red yields violet.

Secondary Financing. Any additional loan or financing that is junior to a mortgage or deed of trust; a Second Mortgage or Second Truct Deed.

Secondary Network. A distribution system wherein the secondary means of an alternating current system are interconnected and supplied through transformers connected in parallel on the secondary side through fuses or automatic switching devices arranged to prevent the feeding of fault current on the primary side of the transformers through the secondary mains; such a system is also called an alternating current automatic low-voltage secondary network.

Secondary Refrigerating System. Refrigerating system in which the condenser is cooled by the evaporator of another or primary refrigerating system.

Second Grade Ceramic Tile. Ceramic tile with appearance defects not affecting wearing or sanitary qualities.

Second Growth. New timber that has grown after the removal of all or a large part of the previous stand.

Second Law of Thermodynamics. Heat will flow only from material at higher temperature to material at a lower temperature.

Second Mortgage. A mortgage recorded after a first mortgage.

Second Trust Deed. A trust deed recorded after a first trust deed.

Section. 1. A drawing showing the kind, arrangement, and proportions, of the various parts of a structure; it shows how the structure would appear if cut through by a plane. 2. Any of several components that could be reassembled to construct an object.

Section Modulus. The ratio of the moment of inertia of a member to the distance from its neutral axis to the outermost fiber.

Section of Land. A land area one mile square, equal to 640 acres; there are 36 sections in a township;

Section Properties. The physical attributes of structural members including dimensions, area, section modulus, moment of inertia, location of centroid and any other geometric information that affects strength.

Security Glass. A glazing sheet with multiple laminations of glass and plastic; designed to stop bullets.

Security Grille. Steel grille to prevent intrusion or entry through openings.

Security Interest. An interest in property, real or personal, held by a creditor to secure the performance of an obligation.

Security System. An interacting or interdirected group of devices, such as alarms and electronic monitors, used to detect breaches of security.

Sedimentary Clay. A clay which has been geologically transported from its place of formation; also called Secondary Clay.

Sedimentary Rock. Rock formed from materials deposited as sediments, such as sand or sea shells, which form sandstone, shale, or limestone.

Seebeck Effect. When two different adjacent metals are heated, an electric current is generated between the metals.

Seepage. Movement of water through soil without formation of definite channels.

Segregation. 1. Varying concentration of concrete ingredients resulting in nonuniform proportions in the concrete mix. 2. The tendency for the coarse particles to separate from the finer particles in handling; in concrete, the coarse aggregate and drier material remains behind and the mortar and wetter material flows ahead; this also occurs in a vertical direction when wet concrete is overvibrated or dropped vertically into the forms, the mortar and wetter material rising to the top. 3. In aggregate, the coarse particles roll to the outside edges of the stockpile.

SEIA. Solar Energy Industries Association.

Seiche. A wave on the surface of water in an enclosed or semi-enclosed area, such as a lake, bay or harbor.

Seismic. Relating to earthquakes.

Seismic Creep. Very slow periodic or episodic movement along a fault trace unaccompanied by earthquakes.

Seismic Design. The structural design of a structure as it is affected by horizontal and vertical stresses caused by earthquakes.

Seismic Hook. In steel reinforcing of reinforced concrete, a 135-degree bend with a six-bar diameter, but not less than 3-inch (76mm) extension that engages the longitudinal reinforcement and projects into the interior of the stirrup or hoop.

Seismicity. The world-wide or local distribution of earthquakes in space and time; a general term for the number of earthquakes in a unit of time, or for relative earthquake activity.

Seismic Load. A load on a structure caused by movement of the earth relative to the structure during an earthquake.

Seismic Wave. One of four distinct waves generated by an earthquake; P-wave, S-Wave, Love wave, and Rayleigh wave.

Seismograph. An instrument which produces a continuous record of earth motion.

Seismology. The science of earthquakes.

Selective Surface. An absorber coating that absorbs most of the sunlight hitting it but emits very little thermal radiation.

Selects. These are the accepted or suitable bricks after culling.

Self-Cleaning. Paint in which the rate of chalking is controlled so dirt on surface will be washed away with accumulated chalk.

Self Furring Lath. Metal lath with dimples that space the lath away from the sheathing behind it to allow plaster to penetrate the lath.

Self Furring. Metal lath or welded wire fabric formed in the manufacturing process to include means by which the material is held away from the supporting surface, thus creating a space for "keying" of the insulating concrete, plaster, or stucco.

Self-Healing. Any a material that melts with solar heat and seals cracks that were formed earlier from other causes.

Self-Inductance. Magnetic field induced in conductor carrying the current.

Self-Leveling. Anyviscous material that is applied by pouring and will spread out prior to curing.

Self Priming. Use of same paint for primer and for subsequent coats; the paint may be thinned differently for the different coats.

Self-Siphonage. The loss of the seal of a trap as a result of removing the water from the trap that is caused by the discharge of the fixture to which the trap is connected.

Self Spacing Tile. Tile with lugs, spacers, or protuberances on the sides that automatically space the tile for the grout joints.

Self Tapping. Creates its own screw threads on the inside of a hole.

Self-Tapping Drywall Screws. Special screws with a drilling type tip configuration for use with metal framing.

Selvage. Also spelled selvedge. 1.The edgeof a sheet material that is finished differently and is expected to be covered in application or cut off. 2. The unsurfaced edge of roll roofing which is covered at the lap when installed. 3. The finished lengthwise edge of woven carpet that will not unravel and will not require binding or serging. 4. Edges of wallpaper without printing.

Semiautomatic Frost Control. Control which starts the defrost part of a cycle manually and then returns the system to normal operation automatically.

Semiconductor. Any of a class of solids, such as silicon or germanium, that conduct no electricity when pure or at low temperatures but are highly conductive when containing suitable impurities or at higher temperatures; used for integrated circuits, transistors, and diodes.

Semi-Drying Oils. Oils which dry to a soft, tacky film; the principal semi-drying oil used in the paint industry is soybean oil.

Semigloss. Sheen on dry finish which is about half way between dead flat finish and full gloss.

Semigloss Enamel. An enamel made so that its coating, when dry, has some luster but is not very glossy.

Semigloss Paint. A paint made so that its coating, when dry, has some luster but is not very glossy.

Semihermetic Compressor. Hermetic compressor with service valves.

Semi-Mat Glaze. 1. A colorless or colored glaze having moderate gloss. 2. A medium-gloss ceramic glaze with or without color.

Semi-Vitreous. Three percent to 7 percent water absorption.

Sensible Heat Gain. The addition of heat to an enclosure by conduction, convection or radiation.

Sensible Heat. Heat which causes a change in temperature of a substance; the degree of molecular excitation of a given mass.

Sensitive Bulb. Part of a sealed fluid device which reacts to temperature; used to measure temperature or to control a mechanism.

Sensor. Material or device which goes through a physical change or an electronic characteristic change as surrounding conditions change.

Sensor, Photoelectric. A device that responds to light and transmits a resulting electrical impulse.

Separate Contractor. A prime contractor; any contractor who has a direct agreement with the owner.

Separation. Division into components or layers by natural causes.

Separation Tank. In a swimming pool, a device used to clarify filter rinse or waste water; also called a Reclamation Tank.

Separator. 1. Device to separate one substance from another. 2. See Interceptor.

Separator, Oil. See Oil Separator.

Septic Tank. A subterranean, usually concrete, tank in which organic sewage matter is decomposed by anaerobic bacterial action.

Septum. That part of the filter element consisting of cloth, wire screen or other porous material on which the filter cake is deposited.

Sequence Controls. Group of devices which act in sequential order.

Serging. A method of finishing a lengthwise cut edge of carpet to prevent unraveling; distinguished from finishing a cut end which may require binding.

Series Circuit. In electrical wiring, the path in which electricity to operate a second lamp or device must pass through the first; current flow travels, serially, through all devices connected together.

Serrate. To punch-perforate in a line to allow for shearing by snapping at the perforation.

Serrated. Notched or toothed on the edge.

Service. The conductors and equipment for delivering energy from the electricity supply system to the wiring system of the premises served.

Serviceable Hermetic. Hermetic unit housing containing motor and compressor assembled by use of bolts or cap screws.

Service Cable. Electrical service conductors made up in the form of a cable.

Service Conductors. The electrical supply conductors which extend from the supply main, duct, or transformers of the serving agency to the service equipment of the premises supplied; for overhead conductors this includes the conductors from the last line pole to the service equipment.

Service Core of Retired Executives (SCORE). A source for limited managerial expertise provided by the Small Business Administration.

Service Disconnect. A switch for, or a means of disconnecting an entire building from electrical service; also called Main Switch.

Service Drop. That portion of overhead service conductors between the pole and the first point of attachment to the building or other structure.

Service Ell. See Street Ell.

Service Entrance Conductors. 1. When overhead, that portion of the service conductors which connect the service drop to the service equipment. 2. When underground, that portion of the service conductors between the terminal box located either inside or outside the building wall, or the point of entrance in the building if no terminal box is installed, and the service equipment.

Service Equipment. The necessary equipment, usually consisting of circuit breaker or switch and fuses, and their accessories, intended to constitute the main control and means of cutoff for the electrical supply to a building or structure.

Service Head. The external fitting used at the juncture of the electrical service to a building and the mast from the breaker panel.

Service Lateral. The underground electrical service conductors between the street main, including any risers at a pole or other structure or from transformers, and the first point of connection to the service entrance conductors in a terminal box inside or outside the building wall; where there is no terminal box, the point of connection shall be considered to be the point of entrance of the service conductors into the building.

Service Load. The live and dead structural loads, without load factors.

Service Manifold. Chamber equipped with gauges and manual valves, used by service technicians to service refrigerating systems.

Service Panel. The main electric panel which houses the main switches and distributes electricity to the branch panels and circuits.

Service Pipe. The water pipe coming into the building from the street main.

Service Raceway. The rigid conduit, electrical metallic tubing, or other raceway, that encloses the electrical service entrance conductors.

Service Sink. A sink with a deep basin to accommodate a scrub pail; used for the filling and emptying of scrub pails, the rinsing of mops, and the disposal of cleaning water; also called a Slop Sink.

Service Station. An establishment that services motor vehicles and sells gasoline and oil.

Service Valve. A type of valve which provides an opening in a liquid piping system to service and monitor the system.

Service Window. An opening in a wall or partition through which business is transacted.

Service Wye. A pipe fitting which joins three pipes at a 45 degree angle; a drainage fitting in a plumbing system.

Servo. 1. A servomotor. 2. A servomechanism.

Servomechanism. A low power device, electrical, hydraulic, or pneumatic, that controls a more powerful mechanism.

Servomotor. The motive element in a servo mechanism.

Set. 1.To harden by chemical hydration. 2. A change from a plastic to a hard state. 3. The change in mortar or plaster from a plastic, workable state to a solid, rigid state. 4. The condition reached by a cement paste, mortar, or concrete when it has lost plasticity to an arbitrary degree, usually measured in terms of resistance to penetration or deformation; initial set refers to first stiffening; final set refers to attainment of significant rigidity. 5. A chisel used for cutting brick.

Set Back. 1. The distance a building is located from the front property line. 2. See Offset.

Set In. The distance a brick is set back from the brick directly below it.

Set Match. In a Set Match carpet pattern, the figure matches straight across on each side of the narrow carpet width; in a Drop Match, the figure matches midway of the design; in a Quarter Drop Match, the figure matches one-quarter of the length of the repeat on the opposite side.

Set Off. See Offset.

Set Screw. 1. A headless screw used to secure two separate parts in a relative position to one another, preventing the independent motion of either part. 2. A screw to adjust the tension of a spring.

Sett. A rectangular paving block of stone or wood.

Setting Bed. 1. The layer of mortar on which the tile is set. 2. The final coat of mortar on a wall or ceiling.

Setting of a Circuit Breaker. The value of the current at which the circuit breaker is set to trip.

Setting Time. The time required for a freshly mixed cement paste, mortar, concrete, or plaster to achieve initial or final set.

Setting Type Joint Compound. A gypsumboard joint compound that hardens by chemical reaction prior to drying; used for patching and completing joint finishing in a shorter period of time.

Setting Up. Initial drying of coating to a point where it is no longer able to flow.

Settlings. Dregs; sediment.

Severability. A rule under which a contract may be divided into its component parts.

Sewage. Liquid waste containing animal, vegetable, or chemical matter in suspension or solution.

Sewage Ejector. A sealed chamber for storing sewage and periodically pumping it up to a sewer at a higher level.

Sewer. A subterranean conduit for conveying sewage or storm water.

Sewer Brick. Low absorption, abrasive-resistant brick intended for use in drainage structures.

Sewer Connection. The juncture where the house drain connects to the public sewer.

Sewer Tile. Glazed waterproof clay pipe with bell joints.

Sewing Pole. Any piece of wood or other material, more or less rounded, over which carpet may be laid prior to opening up the fabric in order to facilitate sewing and other related operations; most carpet layers prefer a wooden pole about 4 in. in diameter that has been slightly flattened on one side.

Sex Bolts. Pairs of bolts, male and female threaded, used for bolting through doors, for example, to attach door closers or surface hinges.

Sexagesimal Measure. The division of a circle into 360 degrees, each degree into 60 minutes, and each minute into 60 seconds.

Sgraffito. A decorative and artistic medium generally consisting of two layers of differently colored plaster; while still soft, the uppermost layer is scratched away, exposing the base or ground layer; countless variations on the process are possible by modulation or pigments and in combination with fresco techniques; sgraffito in Italian means scratched.

Shade. 1. The gradation of color. 2. A color produced by a pigment or mixture with some black in it. 3. A window blind. 4. A device to reduce the glare from a lamp. 5. A screen to provide shelter from light, heat, or sun. 6. A minute variation.

Shaded-Pole Motor. Small a-c motor designed to start under light loads.

Shading. In carpet, bending or crushing the pile surface so that the fibers reflect light unevenly; not a defect but an inherent characteristic of some pile fabrics.

Shading Coefficient. The ratio of total solar heat passing through a given sheet of glass to that passing through a sheet of clear double-strength glass.

Shading Lacquer. Transparent colored lacquer used in shading; applied with a spray gun.

Shadowing. 1. See Joint Photographing. 2. When preceding coats of paint show through the last coat, the finish is said to be shadowing.

Shaft. An unbroken vertical passage through a multistory building, used for elevators, wiring, plumbing, or ductwork.

Shaft Seal. Device used to prevent leakage between shaft and housing.

Shaft Wall. 1. A wall surrounding a shaft. 2. A fire-rated drywall system which is used to enclose a shaft.

Shag Carpet. A carpet made of long coarse or matted fiber; surface consisting of long twisted loops.

Shake. 1. In wood, a separation along the grain, the greater part of which occurs between the rings of annual growth; usually considered to have occurred in the standing tree or during felling. 2. A Handsplit Shingle.

Shake Roof. A roof covering system made up of thicker, hand-cut cedar wood shingles.

Shale. A rock formed of consolidated mud.

Shall. A mandatory term as used in most building laws, construction specifications, and building contracts.

Shared Savings Provision. In a construction contract, when the actual costs including profit and overhead are less than the guaranteed maximum price (GMP), the difference will be shared by contractor and owner on some stipulated percentage basis.

Shark Fin. An upward-curled, felt sidelap or endlap.

Sharp Luster. A very high gloss.

Sharp Sand. Coarse sand of which the particles are of angular shape.

Shear. A force effect that is lateral, perpendicular to the axis of a structure.

Shear Connector. Steel stud or strap, embedded within concrete slab, to transfer horizontal shear loads from slab to supporting structure.

Shear Diagram. A graphic representation of the value of the shear stresses at any point along a beam.

Shearing. A carpet finishing operation that removes stray fibers and fuzz from loop pile, and produces a smooth level surface on cut pile.

Shear Panel. A vertical plane that resists lateral forces; see Shear Wall.

Shear Plate. In heavy timber construction, a round steel plate used for connecting wood to non-wood materials.

Shear Strength. The strength of an element to resist shear.

Shear Stud. 1. A piece of steel welded to the top of a steel beam or girder so as to become embedded in the concrete fill over the beam and cause the beam and the concrete to act as a single structural unit. 2. Forged steel headed stud, fusion welded to horizontal structural decking, embedded within concrete slab, to transfer horizontal shear loads from slab to supporting structure.

Shear Wall. 1. A wall designed to resist lateral forces parallel to the wall. 2. A wall portion of a structural frame intended to resist lateral forces, such as earthquake, wind, and blast, acting in or parallel to the plane of the wall; see Shear Panel.

Sheathing. 1. The rough covering applied to the outside of the roof, wall, or floor framing of a structure. 2. See Gypsum Sheathing.

Sheathing Paper. The water resistant paper installed over sheathing and under siding or shingles to insulate in the house; building paper.

Sheathing, Roof. See Roof Sheathing.

Sheathing, Wall. See Wall Sheathing.

Shed. A building or dormer with a single pitched roof.

Shedding. See Fluffing.

Shed Roof. A single-pitched roof.

Sheen. Special type of gloss measured in terms of reflected light at an angle of 30 degrees or less.

Sheepsfoot Roller. A towed roller consisting of a steel drum with numerous protruding bulbous steel rods, used for compacting clay soil.

Sheet Asphalt. A hot mixture of asphalt cement with clean angular, graded sand and mineral filler; its use ordinarily is confined to a surface course, usually laid on an intermediate or leveling course.

Sheet Flashing. Sheet material, commonly metal, used in construction to direct the flow of water or to prevent water penetration.

Sheet Floor. Any type of resilient flooring material, such as linoleum, that is manufactured and installed in sheets or rolls.

Sheet Flow. Rainwater runoff from land or paving in a continuous sheet in contrast with channeled flow.

Sheet Glass. Window glass.

Sheeting. 7/8" tongue and groove board.

Sheet Lath. A type of metal lath formed by punching geometrical perforations in steel sheets; made from heavier gauge steel than expanded laths, they consequently have greater stiffness.

Sheet Lead. Cold-rolled lead in a sheet whose thickness is expressed by the weight of one square foot of the finished product.

Sheet Metal Brake. A machine for flanging, bending, or folding sheet metal.

Sheet Metal Collar. Sheet metal that covers an object or connection.

Sheet Metal Roof. A roof covering of aluminum, copper, stainless steel, or galvanized metal sheets; in high-finish applications, referred to as an architectural roof.

Sheet Metal Screw. A screw with a fully threaded shank for attaching two thicknesses of sheet metal together; has a pointed tip that will start in a drilled hole and is self-tapping.

Sheet Metal. Flat rolled metal less than 1/4 inches in thickness.

Sheet Piling. Planking or sheeting made of concrete, timber, or steel that is driven in, interlocked or tongue and grooved together to provide a tight wall to resist the lateral pressure of water, adjacent earth or other materials.

Sheet Plastic. Plastic in which thickness is very small in relation to length or width.

Sheetrock. See Gypsum Board.

Sheetrock Beam. A horizontal member constructed of plasterboard.

Sheetrock Finishing. The final sanding and coating of gypsum board seams to make ready for painting.

Shelf. A horizontal mounted surface upon which objects may be stored, supported, or displayed.

Shelf Angle. A steel angle attached to the spandrel of a building to support a masonry facing.

Shelf, Checkroom. See Checkroom Shelf.

Shelf Life. The period of time during which various materials, such as adhesives and sealants, can be stored under specific temperature conditions and remain suitable for use; also called Storage Life.

Shelf Pin. A pin for supporting a shelf, inserted in a hole in a vertical member of a cabinet or bookcase; also called a Shelf Support or Shelf Rest.

Shelf Rest. See Shelf Pin.

Shelf Support. See Shelf Pin.

Shell. A curved, stiff, surface which can carry normal forces by the normal components of tension, compression, or shear forces which can exist within the thickness of the shell.

Shellac. Lac resin melted into thin flakes and used for making varnish; lac is a resinous substance secreted as a protective covering by the lac insect (an Asian scale insect living in trees) and used to make varnish and shellac.

Shell and Tube Flooded Evaporator. Device which flows water through tubes built into a cylindrical evaporator or vice versa.

Shell Concrete. Concrete outside the transverse reinforcement confining the concrete.

Shell Masonry. The outer portion of a masonry unit as placed in masonry.

Shell of Chimney. The outside wall of a chimney.

Shell Type Condenser. Cylinder or receiver which contains condensing water coils or tubes.

Shelter, Golf. See Golf Shelter.

Shelter, Truck. See Truck Shelter.

Shelving. Horizontal mounted surface upon which objects may be stored, supported, or displayed.

Shepherd Violet Toner. A complex manganese phosphate pigment having extreme acid resistance and light fastness.

Sherardizing. A protective coating on metal similar to galvanizing.

SHGF. Solar heat gain through fenestration.

Shield, Expansion. See Expansion Shield.

Shielded Metal-Arc Weld. Welding process where fusion is produced by heating with arc between covered metal electrode and work. Shielding is obtained from decomposition of electrode covering.

Shielding. In welding, a method of protecting adjacent work by positioning temporary protective sheets of rigid material; particularly used for machine applications.

Shim. 1. A thin piece of material placed between two components of a building to adjust their relative positions as they are assembled. 2. To build up low areas; to level or adjust height. 3. To insert shims.

Shingle. 1. A small thin piece of building material often with one end thicker than the other for laying in overlapping rows as a covering for the roof or sides of a building or structure. 2. A small unit of prepared roofing material designed to be installed with similar units in overlapping rows on inclines normally exceeding 25 degrees. 3. To cover with shingles. 4. To apply any sheet material in overlapping rows like shingles.

Shingle Butt. The lower exposed edge of a shingle.

Shingle, Cedar. See Cedar Shingle.

Shingle, Metal. See Metal Shingle.

Shingle Principle. The overlapping of materials, shingle style, so that impinging water, like rainwater, will run harmlessly down and out.

Shingle Roof. 1. Pieces of wedge shaped wood or other material used in overlapping courses to cover a roof. 2. Roof covering of asphalt, asbestos, wood, tile, or slate.

Shingle, Wood. See Wood Shingle.

Shingling Hatchet. A roofer's or sider's shingle hammer which has a small hatchet at its opposite end to shape wood shingles.

Shingling. 1. The pattern that is formed by laying parallel felt rolls on a roof with lapped joints, in which one longitudinal edge overlaps the longitudinal edge of one adjacent felt, and where the other longitudinal edge underlays the other adjacent felt. 2. The application of shingles on a sloped roof.

Ship and Galley Tile. A special quarry tile having an indented pattern on the face of the tile to produce an antislip effect.

Ship Bottom Paint. Special product designed to prevent corrosion and fouling with marine life on the bottom of ships.

Shiplap. 1. A board with edges rabbeted so as to overlap flush from one board to the next. 2. An offset lamination of two layers of gypsumboard.

Shiplapped Lumber. Lumber that is edge-dressed to make a lap joint; see Ship Lap.

Shipping-Dry Lumber. Lumber that is partially dried to prevent stain and mold in transit.

Ships Ladder. Ladder fabricated of metal or wood and inclined between 50° and 77° with steps in place of rungs and provided with railings; usually a standard manufactured item.

Shivering. The splintering which occurs in fired glazes or other ceramic coatings due to critical compressive stress; also called Peeling.

Shoe. 1. A formed metal section used in attaching metal studs to floor and ceiling tracks. 2. The end section of a channel turned to an angle, usually 90 degrees, to permit attachment, generally to other channels.

Shoe Mold. See Base Shoe.

Shooting. See Sprouting.

Shop Coated. Shop Primed.

Shop Drawing Stamp. A rubber imprinting stamp used by architects, engineers, and contractors to indicate their approval, disapproval, or other administrative action on shop drawings.

Shop Drawings. Detailed fabrication or construction drawings of specific items of a project provided by subcontractors.

Shop Lathe. A machine that rotates an object about a horizontal axis which then can be shaped by a fixed cutter or tool.

Shop Painted, Asphalt-Base Type. Coating of asphalt-based paint applied in shop, usually for use in high humidity areas.

Shop Painted, Baked Enamel. Coating of baked enamel applied in shop, usually over a primer on underside of deck where exposed to view.

Shop Painted. Coating of paint applied in shop, usually a primer coat to protect metal from corrosion, which may or may not receive additional coats in the field.

Shopping Center. One or more sales establishments or stores grouped with associated parking facilities.

Shopping Mall. A large suburban building or group of buildings containing various shops with associated passageways.

Shop Primed. Shop painted.

Shore. Temporary support for formwork and fresh concrete that has not developed full strength.

Shore Hardness. The reading of a material's hardness on a durometer similar to the Shore A durometer, the scale of which is 0-100, used on rigid and semi-rigid materials such as polystyrene; consists of a pinpoint depression into the material; both the Shore A and Shore D instruments are made by the Shore Instrument Manufacturing Company, Inc., Jamaica, New York.

Shoring. 1. Temporary bracing to hold the sides of an excavation and prevent it from caving. 2. The timbers used as bracing against a wall or under a beam for temporary support.

Short. Paint that does not have uniform appearance; usually due to the absence of easy brushing liquids.

Short Circuiting. 1. In electrical, an interrupted circuit. 2. In HVAC, a situation that occurs when the supply air flows to exhaust registers before entering the breathing zone; to avoid short-circuiting, the supply air must be delivered at a temperature and velocity that results in mixing throughout the space.

Short Cycling. Refrigerating system that starts and stops more frequently than it should.

Short Term Debt. Mortgages and loans due in less than a year.

Short Time. Short time duty is a requirement of service that demands operation at a substantially constant load for a short and definitely specified time.

Shot. In carpet making, the number of weft yarns in relation to each row of pile tufts crosswise on the loom; a 2-shot fabric is one having two weft yarns for each row of pile tufts; a 3-shot fabric has three weft yarns for each row of tufts; see Weft.

Shot Blasting. Blast cleaning using steel shot as the abrasive.

Shotcrete. A low-slump concrete mixture deposited by being blown from a nozzle at high speed with a stream of compressed air, also called Gunite.

Should. As used in most building laws, construction specifications, and building contracts, it means recommended but not mandatory as when shall is used.

Shoulder. Area between the tapered edge and the face of a gypsum board.

Shoulder Eye Bolt. A bolt type device which has an eye connection as its head.

Shove Joint. Vertical joints filled by mortar by pushing the brick when it is laid in the mortar up against the next brick.

Shove Joints. Vertical joints filled by shoving a brick against the mortar on the next brick when it is being laid in a bed of mortar.

Shovel. A hand implement with a handle and a scoop used for digging, throwing, or lifting material.

Shoving. In asphalt paving, a form of plastic movement resulting in localized bulging of the pavement surface; these distortions usually occur at points where traffic starts and stops, on hills where vehicles brake on the downgrade, on sharp curves, or where vehicles hit a bump and bounce up and down; they occur in asphalt layers that lack stability, usually caused by a mixture that is too rich in asphalt, has too high a proportion of fine aggregate, has coarse or fine aggregate that is too round or too smooth, or has asphalt cement that is too soft; it may also be due to excessive moisture, contamination due to oil spillage, or lack of aeration when placing mixes using liquid asphalts; see Corrugations.

Shower. A bath in which water is showered on the person.

Shower Compartment. An enclosure in which water is showered on a person.

Shower Diverter Valve. A valve at the bath tub filler tap to divert water to the shower over the tub.

Shower Door. A door, usually made of unbreakable plastic or shatterproof glass, used in a shower stall or combination shower/bathtub.

Shower Faucet. A water faucet specifically manufactured for use in a shower.

Shower Floor, Waterproof Membrane. A membrane, usually made of built-up roofing, to provide a positive waterproof floor over the substrate, which is to receive a tile installation using a wire reinforced mortar bed.

Shower Pan. 1. A prefabricated assembly to provide a bottom for a shower, may be made of sheet metal, plastic, or masonry. 2. Terminology used in some areas for a waterproof membrane.

Shower Receptor Liner. Waterproof membrane under the shower floor; also called Shower Receptor Lining.

Shower Receptor Lining. See Shower Receptor Liner.

Shower Receptor. The floor and side walls of the shower up to and including the curb of the shower.

Shower Rod. A horizontal pole onto which a curtain is hung to prevent water from splashing out of a shower.

Shower Stall. A compartment where water is showered on a person.

Show Window. Any window used or designed to be used for the display of goods or advertising material, whether it is fully or partly enclosed or entirely open at the rear, and whether or not it has a platform raised higher than the street floor level.

Shrinkage. 1. In gypsumboard joints, a slight concave depression in the joint treatment usually due to using too thin a mix of joint compound; also called a Starved Joint. 2. The decrease in volume, or contraction, of a material by the escape of any volatile substance, or by a chemical or physical change in the material. 3. Decrease in volume on drying. 4. Loss of bulk of soil when compacted in a fill; usually is computed on the basis of bank measure.

Shrinkage Cracking. Cracking of a structure or member due to failure in tension caused by external or internal restraints as reduction in moisture content develops.

Shrinkage Cracks. In asphalt paving, interconnected cracks forming a series of large blocks usually with sharp corners or angles; frequently they are caused by volume change in either the asphalt mix or in the base or subgrade.

Shrinkage Temperature Steel. Reinforcing bars laid at right angles to the principal bars in a member, for the purpose of preventing excessive cracking caused by drying shrinkage or temperature stresses in concrete.

Shroud. Housing over condenser, evaporator or fan.

Shrub Moving. The digging out of a shrub with its root system intact, to relocate it to a different place.

Shunt. A conductor joining two points of a circuit, through which more or less of a current may be diverted.

Shut-off Valve. A valve used in a supply line to either turn on the liquid or turn it off, seldom used to reduce flow.

SI. French. System International d' Unites; International System of Units.

Sick Building Syndrome (SBS). Describes situations in which building occupants experience acute health or discomfort effects that appear to be linked to time spent in a particular building, but where no specific illness or cause can be identified; the complaints may be localized in a particular room or zone, or may be spread throughout the building.

Side and End Matched. Wood strip flooring that is tongued and grooved on sides and ends.

Side-Grained Wood. Flat-grained lumber.

Side Hill. A slope that crosses the line of work.

Side Jamb. The vertical pieces to a window or door opening on which the door or window rest.

Sidelap. The shortest distance in inches in which horizontally adjacent elements of roofing overlap each other.

Sidelight. A tall, narrow window alongside a door.

Side Lumber. A board from the outer portion of the log. ordinarily one produced when squaring off a log for a tie or timber.

Side Seams. Seam running the length of the carpet; also called a Length Seam.

Side Vent. A vent that connects to the drain pipe through a 45 degree Y or less.

Sidewalk. A walk for pedestrians at the side of a street.

Sidewalk Doors. Steel or aluminum doors in the plane of a sidewalk leading to a basement area.

Side Yard. The space between a building and the side property line.

Siding, Steel. See Steel Siding.

Siding. The outside finish of an exterior wall.

Sienna. A pigment obtained from the earth which is brownish yellow when raw; orange red or reddish brown when burnt.

Sieve. A metallic sheet or plate, woven wire cloth, or similar device, with regularly spaced openings of uniform size, mounted in a suitable frame or holder for use in separating material according to size; also called a Screen; in laboratory work an apparatus in which the apertures are square for separating sizes of material.

Sieve Analysis. Determination of the proportions of particles of the granular material lying within certain size ranges on sieves of different size openings; see Gradation.

Sight Gauge. See Sight Glass.

Sight Glass. A glass section in piping to enable visual monitoring of the condition of the fluid or its height; also called a Sight Gauge.

Sign. A posted command, warning, or direction.

Signage. Any of a group of posted commands, warnings, or directions.

Sign, Electric. See Electric Sign.

Signal Circuit. Any electrical circuit which supplies energy to an appliance which gives a recognizable signal; such circuits include circuits for door bells, buzzers, code-calling systems, and signal lights.

Signboard. A sign.

Silex. A form of silica used extensively in making paste wood fillers; it is chemically inert and does not absorb moisture or shrink.

Silica. SiO_2, the common oxide of silicon usually found naturally as quartz or in complex combination with other elements as silicates; various polymorphs and natural occurrences of silica include cristobalite, tridymite, cryptocrystalline chert, flint, chalcedony, and hydrated opal.

Silica Gel. Hydrated silica in a hard granular form used as a desiccant; when heated, moisture is released and compound may be reused.

Silica Sand. 1. A white fine sand composed of silica. 2. A mineral usually contained in the clay that is used for the making of bricks. 3. An inert pigment made from quartz rock, which is highly resistant to acids, alkalis, heat and light.

Silicate of Soda. See Water Glass, 3.

Silicate Paints. Those employing silicates as binders; used primarily in inorganic zinc rich coating.

Siliceous. Containing silicon or silicate.

Silicon. An abundant non-metallic element occurring in silica and silicates, used in the manufacture of glass, in alloys, and in electronic devices.

Silicon Carbide. An abrasive, shiny black; very hard and brittle crystals made by fusing silica sand and coke in electric furnace.

Silicon-Controlled Rectifier (SCR). Electronic semiconductor which contains silicon.

Silicone. A polymeric substance with high resistance to cold, heat, water, and passage of electricity; used in sealants, lubricants, varnishes, binders, and electric insulators.

Silking. A surface defect characterized by parallel hair-like striations in coated films.

Silk Screen Finishing. Process of finishing where paint is forced through open meshes of a fabric screen; parts of the screen are blocked off and do not print, thus producing the design.

Sill. 1. The lowest member of the frame of a structure, resting on the foundation and supporting the uprights of the frame. 2. The member forming the lower side of an opening, as a door sill or a window sill. 3. See Threshold, 3.

Sill Bolt. Steel bolt placed in the foundation wall or slab to secure the wood sill.

Sill Cock. An outdoor water faucet, usually at sill height, used as a hose connection; also called a Hose Bibb or Wall Hydrant.

Sill Course. See String Course.

Sill High. The height of the window sills.

Sill Pan. A sheet metal deflector installed at each end of a wood window or door sill to carry intruded water that has run down the jamb to the exterior.

Sill Sealer Insulation. Insulation placed between sill plate and supporting concrete or masonry to prevent air leaks.

Silt. 1. Sediment deposited by water in a water course. 2. A soil composed of particles less than 1/20 mm in diameter. 3. A heavy soil intermediate between clay and sand. 4. Particles so fine that they are scarcely visible to the naked eye, unless placed in a glass tube and examined before proper light.

Silt Trap. A settling hole or basin that prevents water-borne soil from entering a pond or drainage system.

Silver. A greyish-white lustrous malleable ductile precious metallic element, used chiefly in alloys; it has the highest thermal and electrical conductivity of any substance.

Silver Brazing. Brazing process in which brazing alloy contains some silver as part of joining alloy.

Silver Leaf. Thin leaf made of silver, used mostly for lettering on glass; aluminum leaf is used where silver color is required on wood or metal surfaces as silver would tarnish.

Simple Beam. A beam that is supported on two supports and where no bending is transferred from the beam to the support.

Sine Wave, A-C Current. Wave form of single frequency alternating current; wave whose displacement is sine of angle proportional to time or distance.

Single Accommodation Sanitary Facility. A room that has not more than one of each type of sanitary fixture, is intended for use by only one person at a time, has no partition around the toilet, and has a door that can be locked on the inside by the room occupant.

Single-Bevel Groove Weld. See Groove Weld.

Single Family Dwelling. A one-family dwelling.

Single Fire. The process of maturing an unfired ceramic body and its glaze in one firing operation.

Single Hung Window. A window with two overlapping sashes, the lower of which can slide vertically in tracks, and the upper of which is fixed.

Single-J Groove Weld. See Groove Weld.

Single Layer. One layer of gypsum board as the finished wall membrane.

Single Lever Faucet. Any of several types of washerless faucets using a single control and springs, balls or cartridges to control flow and temperature.

Single Phase Motor. Electric motor which operates on single- phase alternating current.

Single Ply. A roofing system consisting of only one layer of material.

Single-Pole, Double Throw Switch, (SPDT). Electric switch with one blade and two contact points.

Single-Pole, Single Throw Switch,(SPST). Electric switch with one blade and one contact point.

Single Pole Switch. An electrical switch with one positive contact and two positions, off and on.

Single Roll. A single roll of American-made wallpaper is a roll containing 36 square feet of paper; wallpaper usually comes in bolts which contain two or three single rolls.

Single Spread. See Spread.

Single Stage Compressor. See Compressor, Single Stage.

Single Tee. A precast concrete plank with one integral stiffening rib.

Single-U Groove Weld. See Groove Weld.

Single-Vee Groove Weld. See Groove Weld.

Single Zone Air Handling System. A system supplying conditioned air to an entire building or to a portion of a building that can be controlled as one zone; a constant volume of air is supplied to the space served.

Sinistral. Left handed.

Sink. 1. A basin with a drainage system and water supply, used for washing and drainage. 2. A pit or pool used for the deposit of waste.

Sink, Darkroom. See Darkroom Sink.

Sinking Fund. A fund of money accumulated, and augmented by interest, over time to retire a debt or provide reserves for replacement of wasting assets.

Sintering. Heating a powder until it begins to melt, adding to its strength without destroying its porosity, used in manufacturing lightweight aggregates.

Siphon. A tube used to transfer liquids from a higher to a lower level by means of atmospheric pressure.

Siphonage. A particle vacuum created by the flow of liquids in pipes.

Siren. A device, often electrically operated, for producing a penetrating warning sound.

Sisal. A strong durable white fiber from the fleshy leaves of the agave plant; used in making rope and twine.

Sister Joist. The reinforcement of a joist by nailing, or attaching alongside the existing joist, another joist or reinforcing member.

Site. Area of land to be used for a building location.

Sitecast Concrete. Concrete that is cast-in-place.

Site Demolition. The act or process of demolishing an old building or structure to make way for new construction.

Site Development. On-site and off-site work, including, but not limited to, walks, sidewalks, ramps, curbs, curb ramps, parking facilities, stairs, planting areas, pools, promenades, exterior gathering or assembly areas and raised or depressed paved areas.

Site Plan. A plan drawing showing the site and buildings.

Sitz Bath. See Prerineal Bath.

Size. Solution of gelatin-type material, such as resin, glue or starch, used to fill or seal pores of surface and prevent absorption of finishing materials.

Sizing. 1. Working material to the desired size. 2. A coating of glue, shellac, or other substance applied to a surface to prepare it for painting or other method of finish.

SJI. Steel Joist Institute.

Skarf Joint. An end joint in wood formed by joining with adhesive the ends of two pieces that have been tapered or beveled to form sloping plane surfaces, usually to a featheredge, and with the same slope of the plane with respect to the length of both pieces.

Skein. A batch of yarn.

Skein Dyeing. Immersing skeins of yarn in vats of hot dye.

Sketch. A rough drawing representing the chief features of a site, a building , or an object and often made as a preliminary study.

Skewback. A sloping masonry surface from which a segmental arch springs.

Skid Resistance. 1. A measure of the frictional characteristics of a surface. 2. The ability of an asphalt paving surface, particularly when wet, to offer resistance to slipping or skidding; the factors for obtaining high skid resistance are generally the same as those for obtaining high stability; proper asphalt content and aggregate with a rough surface texture are the greatest contributors; the aggregate must not only have a rough surface texture, but also resist polishing; aggregates containing non-polishing minerals with different wear or abrasion characteristics provide continuous renewal of the pavement's texture, maintaining a skid-resistant surface.

Skim Coat. 1. A thin coat of plaster over any base system; may be the final or finish coat on plaster base. 2. In gypsumboard finishing, a thin coat of joint treatment over the entire surface to reduce surface texture and suction variations. 3. The method of treating plastered areas to receive a gloss paint such as in kitchen and bath areas.

Skimmer. Device that continuously draws swimming pool surface water and surface debris into the filtration system.

Skimmer Weir. Part of swimming pool skimmer that adjusts to small changes in water level and assures continuous flow of water into skimmer.

Skim Filter. A swimming pool surface skimmer combined with a vacuum filter.

Skin Condenser. Condenser using the outer surface of the cabinet as the heat radiating medium.

Skin Friction. The resistance of the soil surrounding a pile to vertical movement of the pile.

Skinning. A tough layer or skin formed on the surface of a paint or varnish in the container; caused by exposure to air.

Skippy. Said of paint that causes the brush to skip on the surface, leaving some spots uncoated and others too thickly coated; this condition can be caused by lack of sufficient vehicle, to permit easy, uniform application or by liquids that pull.

Skips. 1. In painting, holidays, misses, and uncoated spots on finished surface. 2. In roofing, areas that have been missed by the roofer in applying liquid roofing material.

Skip Trowel. A method of plaster texturing resulting in a rough Spanish Stucco effect.

Skirting. Baseboard at a stairway.

Skive Edge. In gypsumboard finishing, the outside edges of the paper joint tape that have been sanded to improve adhesion and reduce waviness.

Skiving. To pare or slice off in thin layers.

Skylight. A transparent opening in a roof for the admittance of light.

Skyscrape. To design or build a very tall building.

Skyscraper. A very tall building; compare with Groundscraper.

Slab. 1. A cast concrete floor. 2. Flat section of floor or roof either on the ground or supported by beams or walls.

Slab Bolster. Continuous, individual support used to hold reinforcing bars in the proper position.

Slab Form. 1. The formwork used for the pouring or placing of a concrete slab. 2. A type of manufactured metal decking which is made expressly to receive a final layer of poured concrete.

Slab Mesh. Welded-wire fabric in sheets or rolls used to reinforce concrete slabs.

Slab, Precast. A flat, horizontal, molded layer of reinforced concrete, cast and cured in other than its final position.

Slack. 1.Extra time in a CPM schedule; also called Float. 2. Hanging loose without tension. 3. Insufficiently diligent; negligent. 4. Slow in business activity.

Slag. A by-product smelting iron, lead, or copper ore; used for construction aggregate; cinder.

Slag Inclusion. Non-metallic solid material entrapped in weld metal or between weld metal and base metal.

Slake. A term denoting the process whereby lime putty is produced from quicklime; slaking consists of adding quicklime to water and allowing the resulting slurry to age for at least two weeks.

Slaked Lime. Hydrated lime.

Slander of Title. Intentionally and wrongfully placing a cloud on title to real estate.

Slash-Grained Wood. Flat-grained lumber.

Slate. A form of geologically hardened clay, easily split into thin sheets.

Slate Flour. Filler used to considerable extent in asphalt mixtures and in roofing mastic.

Sledge. See Sledgehammer.

Sledgehammer. A large heavy hammer that is wielded with both hands; also called a Sledge.

Sleeper. 1. A timber laid on the ground to support a floor joist. 2. A wood strip, usually for a wood floor system, which is fastened directly to a concrete floor thus facilitating the installation of the finished floor.

Sleeping Accommodations. Rooms in which people sleep.

Slenderness Ratio. The ratio of the unbraced length of a column to the radius of gyration.

Sliced Veneer. Wood veneer that is sliced off a log, bolt, or flitch with a knife.

Slicker. A tool often used by the plasterer in place of the darby. It is made of a thin board beveled on both sides, about 4′ long and 6 to 8 inches wide, held by the thicker edge.

Slide. 1. To move in continuous contact with a smooth surface, 2. A fresh tile wall that has buckled or sagged; this condition may be caused by excessive mortar, insufficient lime in the mortar, or excessive moisture in the scratch coat; a slide also may result if the surface is slick or the mortar is too soft. 3. A small landslide.

Slide, Water. See Water Slide.

Sliding Glass Door. An exterior glass door mounted above and below on tracks for ease in movement.

Sliding Window. A window that moves horizontally in tracks.

Sling Psychrometer. Measuring device with wet and dry bulb thermometers; moved rapidly through air it measures humidity; see Hygrometer.

Slip. 1. An inadvertent or trivial mistake. 2. A suspension of ceramic material in liquid. 3. A small geological fault. 4. A long, thin, strip of wood.

Slip Brick. See Soft Mud Brick.

Slip Coating. A ceramic material or mixture other than a glaze, applied to a ceramic body and fired to the maturity required to develop specified characteristics.

Slip Covering. A pipe coupling which has no stop to prevent it from slipping over a pipe; used to make water tight joints in plastic or copper pipe during a repair or alteration.

Slip-Critical Connection. A steel connection in which high-strength bolts clamp the members together with sufficient force that the load is transferred between them by friction.

Slip Form. A form which is raised or pulled as concrete is placed; may move vertically to form walls, stacks, bins or silos, usually of uniform cross section from bottom to top; or a generally horizontal direction to lay concrete evenly for highways, on slopes and inverts of canals, tunnels and siphons.

Slip Forming. Building multistory sitecast concrete walls with forms that rise up the wall as construction progresses.

Slip Glaze. A glaze consisting primarily of a readily fusible clay or silt.

Slip Joint. 1. A contraction joint in a concrete or masonry wall that allows lateral movement. 2. A connection in which one pipe slides inside another allowing for expansion and contraction without breaking.

Slip Nut. A nut used on P straps; a gasket is compressed around the joint by the slip nut to form a watertight seal.

Slippage. The lateral movement between adjacent plies of roofing felt along the bitumen lines resulting in a randomly wrinkled appearance and sometimes exposing the lower plies or even the base sheet to the weather.

Slippage Cracks. In asphalt paving, crescent-shaped cracks that are open in the direction of the thrust of wheels on the pavement surface; they result when there is a lack of good bond between the surface layer and the course beneath.

Slip-Resistant Tile. Tile having greater slip-resistant characteristics due to an abrasive admixture, abrasive particles in the surface or grooves or patterns in the surface.

Slip Sheet. A sheet of paper used in a built up roof installation to allow the roofing to move over the substrate.

Slip Sill. A masonry sill which fits directly into a masonry opening.

Sliver. A Splinter.

Slop Sink (Service Sink). A deeper fixture than an ordinary sink. Custodian's type sink See Service Sink.

Slope. An inclined position or direction; the rate of incline.

Slot. An opening in a member to receive a connection with another part.

Slot, Anchor. A groove in an object into which a fastener or connector is inserted to attach objects together.

Slot Cut. Description of a tile that has been cut to fit around pipes or switch boxes. This tile is usually in the shape of the letter H or the letter L.

Slotted Screwdriver. The most common of screwdriver types, has a flat square blade; also called a Flat Head Screwdriver.

Slot Weld. A weld made in an elongated hole in one member of a lap or tee joint joining that member to that portion of the surface of the other member which is exposed through the hole; the hole may be open at one end and may be partially or completely filled with weld metal; a fillet-welded slot should not be construed as conforming to this definition.

Slough. 1. A swamp or backwater. 2. A creek in a marsh or tidal flat.

Slow Drying. Requiring 24 hours or longer before recoating is possible.

Slow-Curing (SC) Asphalt. Cutback asphalt composed of asphalt cement and oils of low volatility.

Sludge. 1. A muddy, greasy deposit or sediment. 2. Precipitated solid matter produced by water and sewage treatment processes.

Slug. 1. In the English system (feet, pounds, seconds), the slug is that mass which when acted on by a 1 pound force acquires an acceleration of 1 foot per second per second. 2. Detached mass of liquid or oil which causes an impact or hammer in a circulating system.

Slugging. 1. Condition in which a mass of liquid enters a compressor causing hammering. 2. The act of adding a separate piece or pieces of material in a joint before or during welding resulting in a welded joint which does not comply with design, drawing, or specification requirements.

Sluice. 1. A sliding gate or other contrivance for regulating the flow of water in a channel. 2. An artificial water channel. 3. A steep, narrow waterway.

Slump. A measure of consistency of freshly mixed concrete, mortar, or stucco; the decrease in height of wet concrete when a supporting mold is removed; a measure of the consistency of plastic concrete, equal to the number of inches of subsidence of a truncated cone of concrete released immediately after molding in a standard slump cone

Slump Cone. A mold in the form of the lateral surface of the frustum of a cone with a base diameter of 8 inches (203 mm), top diameter 4 inches (102 mm), and height 12 inches (305 mm), used to hold a specimen of freshly mixed concrete for the slump test; a cone 6 inches (152 mm) high is used for tests of freshly mixed mortar and stucco.

Slump Test. The procedure for measuring slump of concrete, mortar, or plaster with a slump cone.

Slurry. 1. A watery mixture of water and any finely divided insoluble material, such as portland cement, slag, or clay in suspension. 2. Cement grout. 3. In making gypsumboard, the gypsum core mixture in its fluid state prior to setting.

Slurry Coat. A brushed application of slurry, generally applied to back of adhered veneer units and to backing.

Slush Coat. A pure coat of a very soft consistency; also called a Slurry coat.

Slushed Joint. Vertical joint filled, after units are laid, by throwing mortar in with the edge of a trowel, used in fireplace and chimney construction; also called a Slush Joint.

Slush Joint. See Slushed Joint.

Small Business Administration. This agency funds small business investment companies and minority small business investment companies that make loans to provide financing for contractors; usually restricted to those with a proven history.

Small Calorie. See Calorie, 1.

Smalt. Deep blue pigment prepared by fusing together potash, silica and oxide of cobalt, and reducing to powder the glass thus formed; smalt is sometimes applied to freshly coated surfaces to provide unusual decorating effect.

Smart Building. A modern office building that is pre-wired for all the telecommunication services and computer links.

Smelt. To melt or fuse ore and to separate metal.

Smelter. A furnace in which the raw materials are melted.

Smoke Alarm. A device that emits a warning sound if it detects smoke.

Smoke Chamber. The space in a fireplace immediately above the throat where the smoke gathers before passing into the flue and narrowed by corbeling to the size of the flue lining above.

Smoke Detector. See Smoke Alarm.

Smoke Developed Rating. An index of the toxic fumes generated by a material as it burns.

Smoke Shelf. The horizontal area behind the damper of a fireplace.

Smoke Test. Test made to determine completeness of combustion.

Smoke Vent. A vent or chimney forming a passageway for expulsion of vent gases from gas-burning units to the outside air.

Smooth-Surfaced Roof. A built-up roofing membrane that is surfaced with a layer of hot-mopped asphalt or cold-applied asphalt-clay emulsion or asphalt cutback; sometimes consisting of an unmopped inorganic felt.

Snake. A flexible tape used by plumbers for clearing stoppages in drain and sewer piping; also called Plumber's Snake.

Snap Header. A false header composed of half bricks with the good end showing on the surface.

Snap Line. See Chalk Line.

Snow Load. The live load of snow accumulating on a roof structure.

Soap Brick. A small brick for filling out a course.

Soap Dish. A concave vessel, sometimes carved into a sink, which holds a bar of soap when not in use.

Soaping Tile. The method of applying a soapy film to newly tiled walls to protect them from paint and plaster during construction.

Soapstone. See Talc.

Social Obsolescence. Depreciation; economic obsolescence.

Socket. 1. An opening to accept a mechanical device. 2. An electrical device to accept a light bulb.

Socket Wrench. A wrench with interchangeable sockets for different sizes of nuts and bolts.

Sod. The grass covered surface of the ground; turf.

Soda Ash. Sodium carbonate is used in swimming pools to raise the pH and increase total alkalinity; it is also used with alum in sand-type filters to produce floc; when chlorine gas is used as a disinfectant, this chemical is used to neutralize hydrochloric acid.

Sodium Bicarbonate. Baking soda used to raise total alkalinity of pool water with little change in pH.

Sodium Bisulfate. Dry acid that, mixed with water, lowers pH and total alkalinity of pool water.

Sodium Fixture. An electric lamp that contains sodium vapor and electrodes between which a luminous discharge takes place, commonly used outdoors.

Sodium Hypchlorite. Liquid containing 12 to 15 percent available chlorine used to disinfect pool water.

Sodium Thiosulfate. A chemical compound used to remove all chlorine from a test sample to avoid false pH test readings or false bacteria test results.

Soffit. 1. The underside part of a member of a structure, such as a beam, stairway, roof, or arch. 2. The undersurface of a horizontal element of a building, especially the underside of a stair or a roof overhang; relatively small in area as compared with a ceiling.

Soffit Board. A special designed and formulated gypsum board for use in exterior over hangs, carport ceilings and other weather protected areas.

Soffit Vent. An opening under the eaves of a roof used to allow air to flow into the attic or the space below the roof sheathing.

Softball Backstop. A three-sided enclosure with an overhang, often made of chain link fencing, that prevents the ball from leaving the field and from hitting spectators.

Soft-Burned. Clay products which have been fired at low

temperature ranges, generally producing relatively high absorptions and low compressive strengths.

Soft Dollars. That portion of equity investment in a real estate development that is tax deductible in the first year, such as prepaid interest.

Softener, Water. See Water Softener.

Softening Point Drift. A lowering of the softening point of bitumen, usually caused by long overheating or mixing asphalt and coal-tar pitch; also called Dropback

Softening Point. 1. The temperature point where a hard material becomes soft or viscous. 2. The softening point of bitumen; the asphalt softening point is measured by the ring and ball test; the softening point of coal-tar pitch is measured by the cube in water test.

Soft Mud Brick. Brick produced by molding relatively wet clay, often a hand process; when insides of molds are sanded to prevent sticking of clay, the product is Sand Mold Brick; when molds are wetted to prevent sticking, the product is Water-Struck Brick or Slip Brick.

Soft Rot. A special type of decay developing under very wet conditions (as in cooling towers and boat timbers) in the outer wood layers, caused by cellulose-destroying microfungi that attack the secondary cell walls and not the intercellular layer.

Soft Water. Water with a minimal mineral content.

Software, Scheduling. See Scheduling Software.

Softwood Plywood Panel. Plywood manufactured from softwood species as Douglas fir, western larch, western hemlock, Sitka spruce, white cedar, or redwood.

Softwoods. 1.Generally, one of the botanical groups of trees that in most cases have needlelike or scalelike leaves, the conifers. 2. The wood produced by such trees. 3. The term has no reference to the actual hardness of the wood.

Soil. The loose surface material of the earth's crust; the unconsolidated natural surface material above bedrock.

Soil and Waste Pipe. Plastic, copper, cast iron or DWV drainage, water and vent.

Soil Bend. A piece of short, curved pipe, like an elbow, used to connect two straight links of pipe in a sewage system.

Soil Cover. A light covering of plastic film, roll roofing, or similar material used over the soil in crawl spaces of buildings to minimize moisture permeation of the area; also called Ground Cover.

Soil Gases. Gases that enter a building from the surrounding ground, such as radon, volatile organics, and pesticides.

Soil, Heavy. See Heavy Soil.

Soil Investigation. See Geotechnical Investigation.

Soil Pipe. A plumbing line that carries waste water from water closets and urinals.

Soil Report. A geological report by a geological engineer providing information on subsurface soil conditions.

Soils Engineer. A professional engineer whose area of expertise is in soil analysis, foundation design, drainage, and the effects of other site conditions on the design of structures.

Soil Stack. A vertical pipe used to carry human waste from toilets; the vertical main piping of a DWV system.

Soil Structure Interaction. The effects of the properties of both soil and structure upon response of the structure.

Soil Test. See Geotechnical Investigation.

Sol-Air Temperature. The combined solar and convective sources of heat at the outside surface of a wall or roof.

Solar Array. An array of panels for collecting heat from the sun for use in a solar heating system.

Solar Collector. See Collector, 2.

Solar Heat Gain through Fenestration. The amount of heat gain through a particular cladding system or window.

Solar Heat. Heat created by visible and invisible energy waves from the sun.

Solar Heating. A heating system utilizing heat from the sun.

Solarium. A glass enclosed porch or room with exposures to the sun, used for relaxation or therapy.

Solar Panel. A battery of solar cells.

Solar Screen Title. Tile manufactured for masonry screen construction.

Solar System. (1) The sun and the celestial bodies whose motion it governs; the sun, with the planets, moons, asteroids, and comets that orbit it. (2) Heating and cooling systems that are based on solar energy.

Solder. A fusible alloy, usually of lead and tin, used to join less fusible metals or wires.

Soldering Iron. A pointed or wedge-shaped tool that is heated electrically or by other means, used in soldering.

Solderless Pressure Connector. A device which establishes the connection between two or more conductors or between one or more conductors and a terminal by means of mechanical pressure and without the use of solder.

Soldier. A brick laid on its end, with its narrow face toward the outside of the wall.

Soldier Course. Oblong tile or brick laid with the long side vertical and all joints in alignment.

Solenoid. A cylindrical coil of wire acting as an electromagnet when a current flows.

Solenoid Valve. A valve that is actuated by an electric solenoid.

Sole Ownership. A business owned entirely by one person.

Sole Plate. The horizontal piece of dimension lumber to which the bottom of the studs are attached in a wall of a light frame building.

Sole Proprietor. The sole owner of a business.

Sole Proprietorship. Same as sole ownership.

Solid. 1. Firm and stable in shape and volume. 2. Having three dimensions. 3. Of the same substance throughout. 4. Without cavities. 5. One of the three states of matter; compare with Liquid and Gas. 6. Nonvolatile portion of paint.

Solid Block. A concrete masonry block with small or no internal cavities.

Solid Casting. Forming castings by introducing a body slip into a porous mold which usually consists of two major sections, one section forming the contour of the inside of the object and allowing a solid cast to form between the two mold faces.

Solid Core Door. A flush door with no internal cavities.

Solid Core. An object with no internal cavities.

Solid Door. A door which is constructed with solid materials.

Solid Masonry Unit. A unit whose net cross-sectional area in every plane parallel to the bearing surface is 75% or more of its gross cross-sectional area measured in the same plane.

Solid Mopping. The mopping of a continuous roof surface area with no areas left unmopped.

Solids. In paint, the dry ingredients remaining after evaporation of all volatile solvent or water; not a fluid and not flowable.

Solid Slab. A concrete slab, without ribs or voids, that spans between beams or bearing walls.

Solids Volume. In paint, percentage of total volume occupied by nonvolatiles.

Solubility. 1. Degree to which a substance may be dissolved. 2. A measure of the purity of an asphalt cement; the portion of the asphalt cement that is soluble in a specified solvent such as trichloroethylene; inert matter, such as salts, free carbon, or non-organic contaminants are insoluble.

Soluble. Describes the property of a substance to dissolve in another and form a solution; for example, sugar is soluble in water.

Soluble Salts. A mineral that is usually present in the clay used for the making of bricks.

Solution. The process by which a substance (solid, liquid, or gas) is homogeneously mixed with a liquid, called the solvent, and the mixture being incapable of mechanical separation into its components; alloys and amalgams are solutions of metals in metal; brines are solutions of a salt in water; syrups are solutions of sugars in water; solution should not be confused or used interchangeably with such terms as dispersion, suspension, or emulsion.

Solution Dyeing. Adding dye or colored pigments to synthetic material while in liquid solution before extrusion into fiber.

Solvent. 1. Having the money to pay all legal debts. 2. Able to dissolve or form a solution.

Solvent Adhesive. An adhesive having a volatile organic liquid as a vehicle, not including water-based adhesives.

Solvent Balance. Ratio of amounts of different solvents in a mixture of solvents.

Solvent Cement. A solvent adhesive that contains a solvent that dissolves or softens the surfaces of plastic pipe being bonded so that the pipe assembly becomes essentially one piece of the same pipe of plastic.

Solvent Release. Ability to permit solvents to evaporate.

Solvent Wash. Cleaning with a solvent.

Sone. Calculated sound loudness rating.

SOP. Standard Operating Procedure.

Sound. 1. Deformation waves that are traveling in the air or other elastic materials. 2. Sound is a rapid fluctuation of air pressure.

Sound Absorbing Material. Acoustic insulation, usually made of glass fiber, placed in flutes of deck to provide increased sound absorption through perforations in deck.

Sound Absorption. 1. The process of dissipating or removing sound energy. 2. The property possessed by materials, objects, and structures such as rooms, of absorbing sound energy.

Sound Absorption Coefficient. The ratio of sound energy absorbed to the sound energy hitting a surface.

Sound Attenuation. A process in which sound is reduced as its energy is converted to motion or heat.

Sound Attenuating Insulation. A special width unfaced mineral fiber product used to improve the sound transmission loss.

Sound Knot. A knot that is solid across its face, at least as hard as the surrounding wood, and shows no indication of decay.

Sound Transmission Class (STC). A single number rating of the sound insulation value of a partition or wall. It is derived from a curve of its insulation value as a function of frequency; the higher the number, the more effective the sound insulation.

Sound Transmission Coefficient. A measure of reduction in sound transmission through certain assemblies of materials used in floors, walls, and ceilings.

Sound Transmission Loss (STL). The decrease or attenuation in sound energy, in decibels, of airborne sound as it passes through a building material; in general, transmission loss increases with frequency, the higher the frequency, the greater the sound transmission loss.

South Pole. 1. The southernmost point of the earth's axis of rotation. 2. The pole of a magnet that points to the south.

Sovereign Immunity. The doctrine that the sovereign is immune to litigation, originally applied to monarchs and now applied to governments.

Soybean Oil. Oil made from seed of soybean, a leguminous annual plant.

Spa. 1. A resort with mineral springs. 2. A unit designed primarily for therapeutic use which is not drained, cleaned, or refilled for each individual; it may include, but is not limited to, hydrojet circulation, hot water, cold water, mineral baths, air induction bubbles, or any combination thereof; industry terminology for spa includes, but is not limited to, therapeutic pool, hydrotherapy pool, whirlpool, or hot spa.

SPA. Southern Forest Products Association.

Space. 1. An area. 2. A distance. 3. A volume. 4. The region beyond the earth's atmosphere and beyond. 5. An amount of area set aside for some purpose, such as storage space or dining space. 6. A definable area, such as a room, toilet room, hall, assembly area, entrance, storage room, alcove, courtyard, or lobby.

Space Dyeing. In carpet making, alternating bands of color applied to yarn by rollers at predetermined intervals prior to tufting.

Space Frame. A truss that spans with two-way action; also called aSpace Truss.

Spacers. T-shaped and Y-shaped, they are used in installation to separate tile on walls an floors. They are manufactured in various thicknesses from 1/16" to 1/2".

Spacer Strip. A metal strip or bar inserted in the root of a joint prepared for a groove weld to serve as a backing and to maintain root opening during welding

Space Truss. See Space Frame.

Spacing Mix. A dry or dampened mixture of one part portland cement and one part extra-fine sand; this mix is used as a filler in the joints of mounted ceramic mosaic tiles to keep them evenly spaced during installation.

Spackle. Trademark; a white powder which, when mixed with water, forms a paste used in filling or repairing cracks in plaster or gypsumboard.

Spackling Compound. Kind of plaster which is used to fill surface irregularities and cracks in plaster; this compound when mixed with paste paint makes what is known as Swedish Putty.

Spading Tool. A thin-bladed shovel which is used to place and tamp concrete.

Spall. A fragment, usually in the shape of a flake, detached from a larger mass by a blow, by the action of weather, by pressure, or by expansion within the larger mass.

Spalling. The cracking, breaking, or splintering of materials, usually due to heat.

Spall, Pile. See Pile Spall.

Span. The distance between supports for a beam, girder, truss, vault, arch or other horizontal structural device.

Spandrel. 1. The wall area between the head of a window on one story and the sill of a window on the floor above. 2. The area of a wall between adjacent arches.

Spandrel Beam. A beam that runs along the outside edge of a floor or roof.

Spandrel Glass. Opaque glass manufactured especially for use in spandrel panels.

Spandrel Panel. A curtain wall panel used in a spandrel.

Spa Pool. A pool, not under medical supervision that incorporates water jets and/or an aeration system used for hydro massage.

Spar Varnish. A very durable varnish designed for severe service on exterior surfaces; resistant to rain, sunlight and heat; so named because of its suitability for the spars of ships.

Spark Arrestor. A mesh grill at the top of a chimney flue to prevent emission of hazardous sparks.

Spark-Proof Tools. Tools made of bronze beryllium.

Spatter. In arc and gas welding, the metal particles expelled during welding and which do not form a part of the weld.

Spatter Finish. Finish which provides a spattered or spackled effect.

Special Conditions. A contract document that revises or clarifies the general conditions, and describes specific conditions of the project or site.

Special Door. A door that has a unique use, such as a bank vault door.

Special-Purpose Tile. A tile, either glazed or unglazed, made to meet or to have specific physical design or appearance characteristics such as size, thickness, shape, color, or decoration; keys or lugs on backs or sides; special resistance to staining, frost, alkalis, acids, thermal shock, physical impact; high coefficient of friction, or electrical properties.

Special Steep Asphalt. A roofing asphalt that has a softening point of approximately 220°F (104°C) and that conforms to the requirements of ASTM Standard D 312, Type IV.

Specialties. A designation of construction materials or components required to finish off the structure; denotes incidental items which are needed to complete construction of the building, such as roof specialties or concrete specialties.

Specialty Contractor. A contractor who follows a recognized trade; trade contractor or subcontractor; a specialty contractor commonly installs certain specific items such as flooring, windows, or terrazzo.

Specification Code. A building code where all allowable and required materials and methods are specified in detail; compare with Performance Cc de.

Specific Gravity. The specific gravity of a substance is its mass (in grams) per cubic centimeter, or it is the weight of a given volume of the substance divided by the weight of an equal volume of water at a temperature of 39.1° F. Hence, one may calculate the specific gravity of any substance by dividing the weight of a cubic foot of the substance (in pounds) by 62.425; as applied to wood, the ratio of the ovendry weight of a sample to the weight of a volume of water equal to the volume of the sample at a specified moisture content (green, air-dry, or ovendry).

Specific Heat. The quantity of heat, in BTUs, needed to raise the temperature of 1 pound of material 1?F.

Specific Humidity. See Humidity Ratio.

Specific Latent Heat of Fusion. The quantity of heat, in joules, required to change unit mass (e.g. 1 kilogram) of a substance from the solid to the liquid state at the melting point.

Specific Latent Heat of Vaporization. The quantity of heat, in joules, required to change unit mass (e.g. 1 kilogram) of a substance from the liquid to the vapor state at the boiling point.

Specific Latent Heat. Latent heat per unit mass.

Specific Performance. 1. The performance of a legal contract substantially in accordance with its terms. 2. A lawsuit to require complete performance of a contract.

Specific Volume. Volume per unit mass of a substance.

Specified Lateral Forces. Lateral forces corresponding to the appropriate distribution of the design base shear force prescribed by the governing code for earthquake-resistant design.

Speckled Glaze. A glaze containing granules of oxides or ceramic stains that are of contrasting colors.

Specks. Any dark dots on the tile less than 1/64 inch in diameter, and noticeable at a distance of more than 3 feet.

Spectral Colors. Band of colors produced when ray of sunshine is bent by glass prism.

Spectrum. The band of colors, as seen in a rainbow, arranged in a progressive series according to their refrangibility or wavelength.

Specular Gloss. Mirror-like reflectance.

Speculation. Assumption of unusual business risk in hopes of obtaining commensurate gain.

Speculum. A reflector in an optical instrument.

Spelter. Zinc, with impurities, cast into slabs for commercial use.

Sphagnum. Atypical mosses that grow only in wet acid areas, combined with other decomposed plant debris to form peat moss.

Sphere. A solid figure, or its surface, with every point on its surface equidistant from its center; a ball shape.

Spheroid. A shape resembling a sphere.

Sphinx. Any of several ancient Egyptian stone figures having a lion's body and a human or animal head.

Spike. A very large nail.

Spike Knot. A knot cut approximately parallel to its long axis so that the exposed section is definitely elongated.

Spillway. An overflow channel for a pond or a terrace channel.

Spinner, Roof. See Roof Spinner.

Spiral. A continuously coiled bar or wire, used for reinforcing in a concrete column.

Spiral Balance. A spring-loaded metallic spiral that provides the counterweight for a doublehung window.

Spiral Column. A concrete column with a continuous spiral of wire around the longitudinal steel.

Spiral, Column. See Column Spiral.

Spiral-Grained Wood. Wood in which the fibers take a spiral course about the trunk of the tree instead of the normal vertical course. The spiral may extend in a right- handed or left-handed direction around the tree trunk; spiral grain is a form of cross grain.

Spiral Reinforcing. Continuously wound reinforcing in the form of a cylindrical helix.

Spiral Spacers. Usually made of channels or angles, punched to form hooks, which are bent over the coiled spiral to maintain it to a definite pitch.

Spire. A tapering cone shaped structure surmounting a church roof or tower; a steeple.

Spirit Level. A device for establishing a horizontal line or plane by means of centering a bubble in a slightly arched glass tube.

Spirit Stain. A stain made by dissolving a dye in an alcohol.

Spirit Varnish. A varnish made by dissolving a resin in a solvent; it dries primarily by evaporation rather than by oxidation.

Spitout. A glaze defect of the pinhole type developed in the decorating kiln, due to evolution of minute gas bubbles from body or glaze.

Splash. 1. A daub or a spot from a splashed liquid. 2. To dash a liquid on a wall or floor. 3. A Splash Wall

Splash Block. A small precast block of concrete or plastic used to divert water at the bottom of a downspout.

Splash System, Oiling. Method of lubricating moving parts by agitating or splashing oil in the crankcase.

Splash Wall. The wall of a tile drainboard or bathtub.

Splay. 1. An inclined surface, as the slope or bevel at the sides of a door or window. 2. To make a beveled surface or to spread out or make oblique.

Splay Angle. Where two surfaces come together forming an angle of more than 90 degrees.

Splice. 1. Joining of two members by overlapping, soldering, or by adhesives. 2. Joining of two similar members in a straight line. 3. Joining two lengths of rope by interweaving the strands. 4. Connection of one reinforcing bar to another by lapping, welding, mechanical couplings or other means. 5. The lap between sheet or rolls of welded wire fabric.

Splice, Compression. See Compression Splice.

Spline. A thin strip of wood used to reinforce joints; also known as a Feather or Tongue.

Spline Joint. A type of connecting strip fitted into slots in each member, used in cabinetmaking.

Splinter. A long, thin, sharp, broken fragment of a material such as wood, metal, or glass; a Sliver.

Split. A tear in the membrane, usually resulting from tensile stress of movement of the substrate; see Crack, 3.

Split Astragal. An astragal that is split through the middle, allowing each door leaf to operate independently.

Split Face Block. A concrete block with a roughened face.

Split Jamb. A door frame fabricated in two interlocking halves, to be installed from the opposite sides of an opening.

Split L Cut. An improper L cut that is made by splitting a tile instead of cutting it.

Split-Phase Motor. Motor with two stator windings; both windings are in use while starting; one is disconnected by centrifugal switch after motor attains speed; motor then operates on other winding only.

Split Receptacle. An electrical outlet where each of the two plugs are on different circuits.

Split Ring. A shear-resisting connector for timbers consisting of a metal ring set in circular grooves in two pieces; the assembly is held together by bolts.

Split Sheet. See Nineteen-Inch Selvage.

Split System. An air conditioning system consisting of two pieces of equipment, an indoor and an outdoor unit; the indoor unit consists of an evaporative coil, provision for heating, and an air handler; the outdoor unit consists of a compressor and an air cooled condenser.

Splitter. A cleavage or lateral fracture in the gypsumboard core; see Core Separation.

Splitting. Tearing or cracking completely through a roof membrane, usually caused by expansion, contraction, deflection, and shearing stresses in the underlying deck construction.

Spodumene (Alpha Spodumene). A lithium mineral of the theoretical composition Li_2O $A_{12}O_34SiO_2$ (monoclinic crystallization) which on heating inverts to beta spodumene, a form having very low thermal expansion.

Spoil. The earth materials that are removed when excavating.

Spokeshave. A hand tool consisting of a blade set between two handles for shaping concave or convex shapes.

Sponge Float. A plasterer's or concrete mason's finishing float that has a sponge surface, for producing a slightly textured finish on the plaster or concrete.

Spool. A reel for winding rope, cord, cable, or wire.

Spool, Insulator. See Insulator Spool.

Spot Grounds. Pieces of wood attached to the plaster base at various intervals for gauging plaster thickness.

Spotlight. A light or lamp which directs a narrow intense beam of light on a small area.

Spot Mopping. 1. In roofing, the application of bitumen in small spots or daubs. 2. An applied mopping pattern in which the hot bitumen is applied in circular areas of about 18 inches in diameter, with a grid of unmopped perpendicular bands.

Spot Repair. Preventive maintenance; repainting of small areas.

Spots. Any dark dots on the face of the tile more than 1/64 inch in diameter.

Spotting. In gypsumboard finishing, to cover fastener heads with joint compound.

Spot Welding. A resistance-welding process wherein coalescence is produced by the heat obtained from resistance to the flow of electric current through the work parts held together under pressure by electrodes; the size and shape of the individually formed welds are limited primarily by the size and contour of the electrodes.

Spot Zoning. The assigning of a particular zoning classification to a parcel of property that is different to the pattern of classification of surrounding property.

Spout. 1. A pipe or conductor through which a liquid is discharged or conveyed in a stream. 2. A pipe carrying rainwater from a roof.

SPR. Simplified Practice Recommendation.

Spray Application. The use of mechanical equipment to apply ceiling or wall texture materials.

Spray Booth. An area in a building or structure used for spray painting; blocked off by walls to prevent dust and dirt from work surface.

Spray Cap. Front enclosure of spray gun equipped with atomizing air holes.

Spray Coating. A coat of paint or other material applied to a surface by use of a spray gun.

Spray Cooling. Method of refrigerating by spraying expendable refrigerant or by spraying refrigerated water.

Sprayed Fireproofing. Fire resistant substances sprayed on construction elements.

Sprayed Insulation. A plastic foam of polyurethane sprayed on a surface to insulate.

Spray Gun. An apparatus resembling a gun for applying a substance such as paint to a surface.

Spray Head. Combination of needle, tip, and air cap.

Spray Loss. See Drift, 2.

Spray Pattern. Configuration of spray when gun is held steady.

Spray Texture. A mechanically applied material used to produce various decorative finishes; may contain aggregates for different effects.

Spray Transfer. A mode of metal transfer in gas metal-arc welding in which the consumable electrode is propelled axially across the arc in small droplets.

Spread. The quantity of paint or adhesive per unit area applied to a surface, usually expressed in pounds of adhesive per thousand square feet of area or square feet per gallon of paint; Single Spread refers to application of adhesive to only one adherent; Double Spread refers to application of adhesive to both adherents.

Spreader. Distribution of poolwater inside filter.

Spreading Rate. Area of surface over which unit volume of paint will spread, usually expressed in square feet per gallon.

Spreadsheet. 1. An outsize tabular sheet used by accountants to distribute figures to diverse accounts. 2. A computer program allowing manipulation and flexible retrieval tabulated numerical data.

SPRI. Single-Ply Roofing Institute.

Spring-Board. Flexible board for competition diving.

Spring Isolator. An isolating device for setting mechanical equipment using coil springs to prevent vibration transmission into the building structure.

Springer. The brick or stone from which an arch springs.

Springer Points. The starting points from which the under curves of an arch start.

Spring-Loaded. A device that contains a compressed or stretched spring pressing one part against another.

Springwood. The portion of the annual growth ring that is formed during the early part of the growing season. It is usually less dense and weaker mechanically than latewood; Earlywood.

Sprinkle Mopping. 1. A throwing of heated bitumen in a random pattern onto the roof surface from a mop or broom, the bitumen consisting of bead size granules. 2. In roofing, the installation of insulation by dipping the mop into hot bitumen and sprinkling it onto the deck.

Sprinkler Head. The outlets from which water is sprayed from a fire sprinkler or irrigation system.

Sprouting. Temporary condition on new carpets where strands of yarn work loose and project above the pile; can be remedied by careful clipping or spot shearing.

Spud. The removal of roofing gravel and heavy accumulations of bitumen by means of chipping and scraping.

Spur Gears. Meshing gears that are on parallel axes.

Square. 1. A tool used by tradesmen to obtain accuracy in laying out right angles. 2. An area measuring 10 feet by 10 feet. 3. An area measure of 100 square feet; usually applied to roofing material; sidewall coverings are often packed to cover 100 square feet and are sold on that basis.

Square Cut. A saw cut perpendicular to the length of a member.

Square Edge. A square factory formed edge; not tapered or otherwise deformed.

Square Fixture. A fixture with four sides of equal length.

Square Footage. The area of a building or parcel of land, expressed in square feet.

Square Groove Weld. See Groove Weld.

Square Manhole. A square vertical access shaft from the surface to underground.

Square Tub. A bathtub in the shape of a square.

Squash Court. A four-walled interior court for playing squash, a game played with balls and racquets.

Squeegee. A rubber edged implement mounted on a handle for spreading, pushing, or wiping liquid material on, across, or off a surface.

Squirrel Cage. Fan which has blades parallel to fan axis and moves air at right angles or perpendicular to fan axis.

SR. Styrene Rubber.

SSD. Saturated Surface Dry

SS Glass. Single strength window glass.

SSPC. Steel Structures Painting Council.

Stability. 1. The ability of a material or structure to remain unchanged. 2. Ability to restore to original condition after being disturbed by some force. 3. The ability of an asphalt paving mixture to resist deformation from imposed loads; stability is dependent upon both internal friction and cohesion.

Stack. 1. Any structure or part thereof which contains a flue or flues for the discharge of gases; a chimney. 2. The vertical main pipe of a soil, waste, or vent pipe system.

Stack Bond. A brick bond consisting of all headers or all stretchers with all vertical joints lined up; also called plumb joint bond, straight stack, jack bond, jack on jack. and checkerboard bond.

Stack Bond. See under Bond.

Stack Effect. Pressure-driven airflow produced by convection as heated air rises, creating a positive pressure area at the top of a building and a negative pressure area at the bottom of a building; the stack effect can overpower the mechanical system and disrupt ventilation and circulation in a building.

Stacking Tile. A method of installation whereby glazed tiles are placed on the wall so that they are in direct contact with the adjacent tiles; the width of the joints is not maintained by the use of string or other means; the tile may be set with either straight or broken joints.

Stack Vent. 1. The continuation of the soil or waste stack, becoming a vent when it is above the highest horizontal or fixture branch that is connected to the stack; siphonage, aspiration, and back pressure are all relieved by air which enters the system through the stack. 2. A vertical outlet in the interior areas of a built- up roof system to provide relief of water vapor trapped in the insulation.

Stadia Rod. A graduated rod used by land surveyors in observing and recording elevations and distances.

Staff. Plaster casts made in moulds and reinforced with fiber; usually wired or nailed into place.

Staff Plaster. A kind of stiff plastering held together with fibrous material, now used for precast plaster moldings.

Staff Room. A room used by staff personnel.

Stage, Band. See Band Stage.

Stage, Theater. See Theater Stage.

Staggered Splices. Splices in steel reinforcing bars which are not made at the same point.

Staggered Studs. Wood wall studs on a plate thicker than the studs with the studs on each side of the wall being staggered, for the purpose of lessening sound transmission from one side of the wall to the other.

Staging. See Scaffold, 1.

Stain. 1. A defective discoloration. 2. A discoloration in wood that may be caused by such diverse agencies as micro- organisms, metals, or chemicals. 3. Materials used to impart color to wood. 4.

Stained Glass. 1. Glass colored or stained for use in windows. 2. Colored glass in a lead framework.

Staining. 1. Discoloration caused by a foreign matter chemically affecting the material itself. Application of a paint which lets the grain of wood show through.

Stainless Grab Bar. A length of metal used as a handrail, constructed of stainless steel, attached to a wall in a bathroom, in a shower, above a bathtub or near a water closet; aids people in the use of bathroom fixtures.

Stainless Steel. An alloy of steel with chromium and other metals that is practically immune to rusting and ordinary corrosion

Stainless Tile. A thin square or rectangular unit used as a finish for walls, floors or roofs, constructed of stainless steel.

Stain Wax, Penetrating. See Penetrating Stain Wax.

Stain, Wood. See Wood Stain.

Stair Formwork. The support for freshly poured or placed concrete for a stair system.

Stair Horse. See Carriage.

Stair Landing. The level platform between flights of stairs.

Stair Pan. A stair assembly constructed to hold precast or cast-in place masonry or stone in sheet metal pans at the treads.

Stair Rail. A handrail at a stairway.

Stair Reinforcing. The reinforcing steel used in the construction of a concrete stair.

Stair Stringer. One of the structural supporting members of a stairway.

Stair, Access. A stair system to provide specific access to roofs and mechanical equipment rooms.

Stair, Concrete. A stair system constructed from concrete.

Stair, Marble. A stair system constructed of or faced with marble.

Stair, Metal. A single or series of metal steps connected by landings.

Stairs, Box. Stairs built between walls, and usually with no support except the wall.

Stair, Terrazzo. A stairway with terrazzo as the wearing surface.

Stair Tread. The horizontal surface of a step in a stairway.

Stair Tread, Ceramic Tile. Ceramic tile installed on the horizontal surface of a stairway.

Stair Tread, Metal-Pan Concrete Filled. Section formed from sheet metal to receive concrete fill which will function as a stair tread.

Stair Tread, Steel Grating. Tread fabricated from metal grating.

Stair Tread, Steel Plate. Tread fabricated of floor plate.

Stairway. One or more flights of stairs including the landings leading from one floor to another.

Stairwork. A unit of horizontal treads, vertical risers, that make up a system of steps from one level to another.

Stake. 1. A stout stick or post sharpened at one end and driven into the earth as a support or boundary marker. 2. Something that is invested for gain or loss; an interest or share in a financial undertaking. 3. A Mormon territorial jurisdiction comprising a group of wards.

Stake, Side. On a road job, a stake on the line of the outer edge of the proposed pavement; any stake not on the center line.

Stake, Slope. In earthwork, a stake marking the line where a cut or fill meets the original grade.

Stall. 1. A booth or counter in a market or building for the sale of goods. 2. A compartment for one domestic animal in a stable. 3. A compartment for one person in a shower bath; a stall shower. 4. A space marked off for parking one car.

Stall Torque. In an electric motor, the torque developed when starting.

Stall, Shower. A compartment where water is showered on a person.

Stanchion. An upright, vertical support or strut.

Standard. 1. A stanchion or support. 2. A criterion accepted by custom or general agreement. 3. A level of quality or quantity established by authority for size, weight, value, or extent.

Standard Air. See Standard Conditions.

Standard Conditions. Used as a basis for air conditioning calculations, temperature of 68°F. (20°C.), pressure of 29.92 inches of mercury and relative humidity of 30 percent; also called Standard Air.

Standard Dimension. The specified dimension of units.

Standard Dimension Ratio (SDR). The ratio of pipe or tubing diameter to wall thickness.

Standard Grade Tile. Highest grade of all types of ceramic tile.

Standardize. To bring into conformity with a standard.

Standard of Living. The degree of comfort available to people.

Standard Operating Procedure. A standard or prescribed method to be followed routinely for the performance of designated operations or in designated situations.

Standing Finish. The finish woodwork of the openings and the base, and all of the interior finish work.

Standing Seam Roof. A sheet metal roof system that has seams that project at right angles to the plane of the roof.

Stand Oil. Heat-thickened vegetable oil, or combination of vegetable oils, such as linseed and tung.

Standpipe. A pipe that extends the full height of a building, with hose connections, used to provide water exclusively for the fighting of fires; see Dry Standpipe and Wet Standpipe.

Stand Pipe System. A system of piping installed for fire protection purposes having a primary water supply constantly or automatically available at each hose outlet.

Stanley Knife. Trademark; a hand tool with removable blades; see Board Knife.

Stannous. Containing or relating to tin.

Staple. Double pointed, U-shaped metal fastener used for same purposes as nails, but providing additional head holding power.

Staple Fibers. Relatively short natural or synthetic fibers ranging from approximately 1-1/2 to 7 inches in length, which are spun into yarn, as contrasted with continuous filament.

Staple Gun. A spring-loaded or pneumatic devise for driving staples.

Star Drill. 1. A drill bit for cutting holes into concrete. 2. A cold chisel, driven by hammer, with a cross-shaped head for cutting holes in concrete.

Starch. An additive in the core that improves bond between the core and facing paper of gypsumboard.

Starch Coating. Protective coating for surfaces coated with flat paint; it also can be used on wallpapers made with colors that do not smear when wet; the coating is made by soaking ordinary laundry starch in a small quantity of cold water to break up lumps; boiled water is then poured on to cook the starch and make it transparent; mix to consistency of cream, let cool, and apply with large paint or calcimine brush; coating should be stippled while still wet to remove brush marks; it may be removed later by using water and sponge.

Starter. A device that insures that a motor does not receive too high a current when starting up.

Starter, Fluorescent Lamp. See Ballast, 1.

Starter, Magnetic. See Magnetic Starter.

Starter Strip. A thin strip of wood used to begin the first course of a horizontal siding system.

Starting Relay. Electrical device which connects or disconnects starting winding of electric motor.

Starting Torque. In an electric motor, the amount of torque available, when at 0 speed, to start and accelerate the load.

Starting Winding. Winding in electric motor used only briefly while motor is starting.

Starved Joint. 1. A glue joint that is poorly bonded because an insufficient quantity of adhesive remained in the joint. 2. See Shrinkage 1.

Statement of Changes in Financial Position. A financial schedule that shows how funds were obtained and where funds were used; also called Statement of Sources and Application of Funds.

Statement of Sources and Application of Funds. See Statement of Changes in Financial Position.

Stateroom. A private passenger compartment on a ship or train.

Static. 1. Stationary, not acting or changing; passive. 2. Concerned with bodies at rest or forces in equilibrium. 3. In spot, seam and projection welding, the force between the electrodes under welding conditions but with no current flowing and no movement in the welding machine.

Statical Moment. The product of a given area and the distance from its centroid to a point of rotation.

Statically Determinant Structure. A structural frame in which the bending moments and reactions can be determined by the laws of statics alone.

Statically Indeterminant Structure. A structural frame in which the bending moments and reactions cannot be calculated from the equations of statics.

Static Bending. Bending under a constant or slowly applied load; flexure.

Static Force. Static Load.

Static Friction. The tangential surface resistance between two bodies in contact which move or tend to move with respect to each other.

Static Head. Pressure of fluid expressed in terms of height of column of the fluid, such as water or mercury.

Static Load. A non-moving load imposed on a structure.

Static Pressure. Condition that exists when an equal amount of air is supplied to and exhausted from a space; at static pressure, equilibrium has been reached.

Statics. The science of bodies at rest or forces in equilibrium.

Station Yards of Haul. In hauling excavated earth or fill, the number of cubic yards material multiplied by the number of 100 foot stations through which it is moved.

Stationary. Remaining in one place; not portable; fixed in position.

Stationary Blade Compressor. Rotary pump which uses a non-rotating blade inside pump to separate intake chamber from exhaust chamber.

Stator. A stationary part in a machine in or about which a rotor revolves.

Stator, Motor. See Motor Stator.

Statute of Frauds. A statute that requires certain contracts to be evidenced by a writing.

Statute of Limitations. The period of time after a cause of action arises before the expiration of which a plaintiff must file suit, or lose the right to do so.

Statute of Repose. A statute of limitations, measured from the completion of a work of improvement, protecting builders, contractors, architects, and engineers from liability for defects for an indefinite time in the construction of the improvement.

Stay-Tackling. Temporary tacking of the carpet in stages during a long stretch with the power stretcher or knee kicker, to hold the stretch until the end of the carpet can be kicked over the pins.

STC. 1. Sound Transmission Coefficient. 2. Sound Transmission Class.

Std. Standard.

Steam. The invisible gaseous state of water, produced by boiling.

Steam Bending. The process of forming curved wood members by steaming or boiling the wood and bending it to a desired shape.

Steam Boiler. A boiler for producing steam.

Steam Clean. The act or process of cleaning a structure with a machine that provides pressurized steam through a nozzle.

Steam Curing. Curing of concrete or mortar in water vapor at atmospheric or higher pressures and at temperatures between about 100° F and 420° F (40° C and 215° C).

Steam Distilled Wood Turpentine. Turpentine made from pine tree stumps by treating shredded chips with live steam to produce a distillate which is fractionated to yield turpentine, pine oil, and solid residue.

Steamer, Electric. See Electric Steamer.

Steam Heating. Heating system in which steam from a boiler is piped to radiators in space to be heated.

Steam Jet Refrigeration. Refrigerating system which uses a steam venturi to create high vacuum (low pressure) on a water container causing water to evaporate at low temperature.

Steam Kettle. A metallic vessel for creating and holding steam, used in a commercial kitchen.

Steam Meter. An instrument for measuring the passage of steam.

Steam Sterilizer. A machine using steam to sterilize implements.

Steam Table. See Bain-Marie

Steam Trap. A device that allows the passage of a condensate or air, but prevents the passage of steam.

Steatite Porcelain. A vitreous ceramic whiteware for technical application in which magnesium metasilicate (MgO, SiO_2) is the essential crystalline phase.

Steatite Talc. Massive talc or the pulverized product, thereof having the general formula 3MgO 4Si O_2H_2O.

Steel. Iron compounded with other metals to increase strength and wearing or rust resistance.

Steel Access Door. Fabricated small steel door and frame, usually flush with adjoining surface to provide access to concealed equipment or system components for inspection and maintenance purposes.

Steel Access Hole Cover. Sheet metal used to cover access hole in deck to facilitate welding operation.

Steel Access Panel. Removable panel, usually flush with adjoining surface to provide access to concealed equipment or system components for inspection and maintenance purposes.

Steel Adjusting Plate. Sheet steel or segments of deck used to fill narrow areas where regular deck units cannot be accommodated.

Steel Anchor Bolt, Foundation. Bolts that tie sill plate and frame of structure to its foundation.

Steel Anchor Bolt, High-Strength. High strength steel bolt used to secure wood construction to concrete or masonry construction.

Steel Anchor Bolt. Steel bolt used to secure wood construction to concrete or masonry construction.

Steel Angle Lintel. Lintel made of structural steel angle.

Steel Angle Sill. Steel angle in concrete or masonry used as a sill.

Steel Angle. An L-shaped member constructed of steel, often used as a lintel or carrying shelf for masonry.

Steel Base Plate. Foundation plate of metal on which heavy piece of machinery or column rests; plate is usually set on masonry or concrete.

Steel Bearing Plate. Flat steel plate attached to supporting building component to uniformly transmit loads to that supporting component.

Steel Bolt. Metallic pin or rod having a head at one end and threads to receive a nut on the other used for holding members or parts of members together; head may be round, square, flat, oval, hexagonal, or other shapes.

Steel Bolt, Common Grade. Bolt made of carbon grade steel for use in ordinary bolted connections in contrast to high strength bolted connections.

Steel Bolt, High-Strength. Quenched and tempered or heat-treated steel bolts with higher tensile strengths.

Steel Bolt, Interference Body. High strength bolt with hardened steel ribs on the bolt shank to facilitate driving of bolts in misaligned holes.

Steel Bolt, Load Indicator. High strength bolt with manufactured notch between bolt tip and threads designed to react to opposing rotational torques applied by installation wrench so bolt tip automatically shears off when proper tension of bolt is achieved.

Steel Bolt, Stainless. Stainless steel bolt used in ordinary bolted connections in contrast to high strength bolted connections.

Steel Bolt, Tamperproof. Steel bolt with a head that prevents bolt from being removable.

Steel Bolt, Weathering. Bolt made of steel to be compatible with weathering steel.

Steel Box Beam. A hollow built-up horizontal structural member.

Steel Cant Strip. Sloped sheet metal strip used as perimeter of roofing membrane to transition membrane from horizontal to vertical surface.

Steel Closure Plate, Deck. Sheet metal plate used to close openings formed between flutes of composite deck and abutting walls or partitions.

Steel Cofferdam. A watertight steel enclosure from which water is pumped to expose the bottom of a body of water and permit foundation construction.

Steel Column. Long relatively slender vertical load bearing steel member.

Steel Column Closure. Sheet metal used to close openings around structural steel penetrations of deck.

Steel Conduit. A pipe, tube, or channel used to enclose electric wires or direct the flow of a fluid.

Steel Control Joint Cover. Exposed linear steel device to cover and conceal a control-joint.

Steel Corner Guard. Steel covering over lower portion of exterior corner of intersection of walls or at corner of column to protect from damage.

Steel Cover Plate and Frame. Sheet steel used to cover non-lapping end joints of deck.

Steel Crane Rail, Overhead, Top Running. Special structural steel shape to support rolling crane.

Steel Embedment Angle. Steel angle embedded in concrete or masonry; used to anchor or support other materials.

Steel Embedment Plate. Rolled steel plate embedded in concrete or masonry; used to anchor or support other materials.

Steel End Closure. Formed sheet metal at 90 degrees to deck profile to contain wet concrete or gypsum fill during placement at slab edges.

Steel Eye Bolt. Steel bolt comprised of a threaded shank with a lopped head designed to accept a hook, cable, or rope.

Steel Fire Escape. Continuous unobstructed route of escape from building in case of fire constructed of steel components, usually located and mounted on exterior of building and composed of stairs, ladders, and landings.

Steel Float. A concrete finisher's tool used to apply the final finish on a concrete slab, also used to apply stucco or plaster.

Steel Floor Deck, Cellular. Steel floor deck which during construction supports wet concrete and construction loads, but after concrete cures does not necessarily perform a structural function in completed construction; deck is fabricated of two sheets to form linear voids which may be used as raceways.

Steel Floor Deck, Cellular, Wire Raceway. Steel floor deck fabricated of two sheets to form linear voids and serve as a raceway for conductors.

Steel Floor Deck, Composite. Steel floor deck which during construction supports wet concrete and construction loads, but after concrete cures and reaches design strength, acts together with concrete to support dead and live loads; deck and concrete are interlocked by shape of deck.

Steel Floor Deck, Form. Steel floor deck which during construction supports wet concrete and construction loads, but after concrete cures and reaches design strength, does not perform structural function in completed construction.

Steel Floor Deck. Sheet steel formed to fluted or ribbed profile to span between supports to support floor system and live loads.

Steel Floor Deck, Non-Cellular Form. Steel floor deck which during construction supports wet concrete and construction loads, but after concrete cures and reaches design strength, does not perform a structural function in completed construction.

Steel Floor Plate. Steel plate usually with raised pattern to provide nonslip wearing surface.

Steel Floor Plate, Patterned. Steel plate with raised pattern wearing surface.

Steel Floor Plate, Raised Tread. Steel plate with raised tread wearing surface.

Steel Forms. A temporary structure for the support of concrete while setting.

Steel Frame. 1. Frame made of steel. 2. A structural steel framed structure.

Steel Grate. A screen made from sets of parallel steel bars placed at right angles to each other, and which allows water to drain.

Steel Grid Walkway. Walkway fabricated of steel grid placed over roof surface to protect roofing surface from damage from traffic.

Steel Gutter. A shallow channel constructed of galvanized sheet steel, positioned just below and along the eaves of a building for collecting and diverting water from a roof.

Steel Hatch. An opening in a floor or roof having a hinge or completely removable cover constructed of steel.

Steel Joist. Open web, parallel chord, load-carrying members suitable for direct support of floors and roof decks, utilizing hot rolled or cold formed steel.

Steel Joist Bridging. Diagonal or longitudinal structural steel members used to keep joists properly spaced, in lateral position, vertically plumb, in order to distribute load.

Steel Joist, DLH Series. Long span steel joist of deeper and heavier construction for longer spans.

Steel Joist Girder. Steel joist of deeper and heavier construction used as primary framing members to support other structural members at points along its span.

Steel Joist, K Series. Open web, parallel chord, load-carrying members suitable for direct support of floors and roof decks, utilizing hot rolled or cold formed steel.

Steel Joist, LH Series. Steel joist of deeper and heavier construction for longer spans.

Steel Ladder Cage. Open steel framework enclosing ladder on open side to prevent falls from ladder; commonly required by safety codes on high ladders.

Steel Lag Screw. Threaded wood screw with square head, to be turned by wrench; usually longer and heavier than common wood screw with coarser thread and larger head for heavier wood construction.

Steel Lintel. 1. In steel construction, a steel member over an opening to carry loads above the opening. 2. A steel member placed within a masonry wall or partition to support masonry or other construction over the opening.

Steel Lock Washer. Split steel washer which compresses when bolt or screw is tightened to prevent loosening of tightened nut or screw.

Steel Nut, Common Grade. Nut made of carbon grade steel•for use with carbon grade bolts in contrast to high strength bolted connections.

Steel Nut, High-Strength. Nut made of alloys for use with high strength bolts.

Steel Nut, Self-Locking. Nut with locking pin that slides along bolt threads. Reversing direction of locking allows nut to be removed without damaging nut or bolt.

Steel Nut, Weathering. Nut made of steel to be compatible with weathering steel.

Steel Pile. A long, slender, piece of steel driven into the ground to act as a foundation.

Steel Pin and Roller. Joint used in steel bridge construction to allow for rotation and movement.

Steel Pipe, Seamless. S eamless steel pipe manufactured by a spinning process.

Steel Plate. Sheet steel of a heavier thickness.

Steel Plate Column Wainscot. Steel covering over lower portion of column to protect it from damage.

Steel Pole. A long steel pole installed vertically in the ground.

Steel Rail Standard. Vertical component of railing system supporting handrail.

Steel Repair. The act or process of restoring steel.

Steel Ridge Plate. Sheet metal plate, placed over ridge formed when two differing roof slopes intersect.

Steel Rivet. Metallic pin or rod having head at one end which is inserted through holes in materials to be fastened together; protruding end is flattened (mushroomed) to tie two pieces together.

Steel Roof Deck. Sheet steel formed to fluted or ribbed profile to span between supports to support roofing system and live loads.

Steel Roof Deck Diaphragm Action. In-plane action of roof deck system such that columns framing into roof from below and above are maintained in their same position relative to each other.

Steel Roof Deck, Acoustical. Steel roof deck with vertical faces perforated to increase room sound absorption properties and flutes filled with sound absorbing material.

Steel Roof Deck, Cellular. Sheet steel formed to fluted or ribbed profile to span between supports to support roofing system and live loads; deck is fabricated of two sheets to form linear voids which may be used to contain electrical conductors.

Steel Roof Deck, Composite. Steel roof deck which during construction supports wet concrete and construction loads, but after concrete cures, acts together with concrete to support dead and live loads; deck and concrete are interlocked by shape of deck or by mechanical means.

Steel Roof Deck, Form. Steel roof deck which during construction supports wet concrete and construction loads.

Steel Roof Deck, Non-Cellular. Steel roof deck which during construction supports wet concrete loads, but after concrete cures and reaches design strength, does not perform structural function on completed construction.

Steel Roof Deck, Vented. Steel deck with perforations or slots to aid in curing and exiting of moisture from concrete or gypsum fill placed on deck.

Steel Roof Panel. A rectangular sheet of steel used in roof covering applications.

Steel Roof Walkway. Fabricated steel walkway used over roof surfaces.

Steel Screw. Steel pin or rod having head at one end and helically threaded with pointed end on other for penetrating material by being turned used for holding members or parts of members together; head may be round, square, flat, oval, hexagonal, or other shapes.

Steel Screw, Self-Tapping. Type of screw which drills its own hole through two pieces to be joined, then taps a thread in both pieces while fastening them together.

Steel Screw, Tamperproof. Steel screw with a head that prevents removal.

Steel Sheet Piling. A steel plank in close contact or interlocking with other steel planks to provide a tight wall to resist the lateral pressure of water, adjacent earth, or other materials at excavations.

Steel Shingle. A roof covering or wall finishing unit made of steel applied on roof systems or exterior wall coverings.

Steel Ships Ladder. Ladder fabricated of steel inclined between 50 degrees and 77 degrees with steps in place of rungs and provided with railings.

Steel Siding. Formed and finished steel panels used as exterior wall coverings.

Steel Sill Angle. Steel angle used as a sill.

Steel Square. A carpenter tool which establishes a right angle between two components; the large arm of the square is called the Body or Blade; the smaller arm, at a 90 degree angle to the blade, is called the Tongue; the point where the outside edges of the blade and tongue join is called the Heel; the surface with the manufacturer's name is called the Face; the opposite surface is called the Back.

Steel Stair. Flight or series of connected stair flights fabricated of steel members extending between two or more floors within given floor area.

Steel Stair, Spiral. Stair of steel construction with a closed circular plan, uniform sector shaped treads and supporting center column.

Steel Strap. A steel plate fastened across the inner section of two or more timbers.

Steel Stringer. Steel member upon which stair treads bear.

Steel, Structural. Steel that is rolled in a variety of shapes and manufactured for use as structural load-bearing members.

Steel Stud. 1. Steel pin or rod having head at one end for driving into material used for holding members or parts of members together. 2. In light gage construction, a vertical framing member similar to a 2x 4 stud.

Steel Sump Pan. Sheet metal pan forming low point in roof deck to collect water and receive roof drain.

Steel Tank. A receptacle manufactured from steel for holding, transporting or storing liquids.

Steel Toggle Bolt. Steel bolt and nut assembly to fasten objects to hollow construction; assembly from only one side; nut has pivoted wings that close against spring when nut end of assembly is pushed through hole and spring open on other side in void of construction assembly.

Steel Toilet Partition Support Framing. Framing, usually above a ceiling used to support ceiling-hung toilet partitions.

Steel Transom. A steel transverse piece in a structure, lintel; a steel horizontal crossbar in a window, over a door, or between a door and a window or fanlight above it.

Steel Valley Plate. Sheet metal plate, placed over valley formed when two differing roof slopes intersect.

Steel Wainscot. Steel covering over lower portion of wall or partition to protect from damage.

Steel Washer. Flat ring of steel which may be plain split, toothed, or embossed, used in threaded connections to distribute loads, span large openings, relieve friction, or prevent loosening.

Steel Washer, High-Strength. Flat ring of steel which may be plain split, toothed, or embossed, used in threaded connections to distribute loads, span large openings, relieve friction, or prevent loosening, made of high strength steel; often used with high strength bolts.

Steel Washer, Load Indicator. Compressible washers with direct tension indicator capable of indicating achievement of minimum bolt tension of high strength bolts.

Steel Washer, Weathering. Washer made of steel to be compatible with weathering steel.

Steel Washer, Welding. Washer made of steel to be used for welding in steel fabrication.

Steel Welding. A localized heating of two pieces of steel to melting until there is a fusing of the material.

Steel Welding Washer. Washer for containing arc spot welds on thin sheet metal (thinner than 0.028 inch).

Steel Wood Screw. Steel screw with threads designed for penetration of wood and similar materials.

Steel Wool. An abrasive material consisting of a mass of fine steel shavings, used for polishing and burnishing.

Steep Asphalt. A roofing asphalt that has a softening point of approximately 190°F (88°C) and that conforms to the requirements of ASTM Standard D 312, Type III.

Steeple. A spire, as on a church.

Steeplejack. A person who climbs tall structures, usually with safety rigging, for painting and repairs.

Steep Slope Roof. Roofing that is over 1-1/2 inches per foot slope or pitch.

Stem. The vertical part of a concrete or masonry retaining wall.

Stenciling. Placing a design on a wall or other surface by applying the finish through a template cut out of thin, flat paper or metal.

Step Flashing. Flashing and counterflashing installed where a sloping roof meets a vertical wall, as at a chimney.

Step Ladder. A short free-standing ladder with flat steps.

Stepped Foundation. A benched foundation on a sloping site with all bottoms flat to prevent sliding.

Stepping Stone. 1.A raised stone used in crossing a stream. 2. Spaced flat stones used as a walk in a garden.

Step Return. A bottom step, the nosing of which extends out considerably over the riser; it is frequently wider than the rest of the stairs, and is usually curved at the ends; also called a Bullnose.

Steri Lamp. Lamp with a high-intensity ultraviolet ray used to kill bacteria; used in food storage cabinets and in air ducts.

Sterilizer. One of three groups of antimicrobials registered by EPA for public health uses; EPA considers an antimicrobial to be a sterilizer when it destroys or eliminates all forms of bacteria, fungi, viruses, and their spores; because spores are considered the most difficult form of a microorganism to destroy, EPA considers the term sporicide to be synonymous with sterilizer.

Sterilizer, Medical. See Medical Sterilizer.

Stethoscope. Instrument used to detect sounds and locate the origin.

Stick. 1. A woody piece of twig or branch of a tree. 2. A slender piece of wood. 3. To pierce with a pointed instrument. 4. Adhesive quality.

Stick-Built. Construction of a wood framed building piece by piece on-site as contrasted to prefabrication.

Stick, Hook. See Hook Stick.

Sticker. A piece of metal channel inserted in concrete or masonry walls for the attachment or support of wall furring channels.

Sticker Stain. A brown or blue stain that develops in seasoning lumber where it has been in contact with the stickers.

Stickers. Strips or boards used to separate the layers of lumber in a pile and thus improve air circulation.

Stick Shellac. Shellac which comes in sold stick form; used extensively for furniture patching.

Stickyback. A transparent decal containing technical information that can be adhered to a tracing.

Stiffback. A structural member placed on top of ceiling joists at midspan to line up the joists before plastering or drywall installation.

Stiffener. A horizontal metal shape tied to vertical members (studs or channels) of partitions or walls to brace them.

Stiffener Plate. A steel plate attached to a structural member to support it against heavy localized loadings or stresses.

Stiffness. The quality of resistance to deformation on the part of a material, a component member of a structure, or the whole structure.

Stiff Mud Brick. Brick produced by extruding a stiff but plastic clay through a die.

Stile. A vertical framing member in a panel door.

Still Water. A part of a stream where the gradient is so gentle that no current is visible.

Stillson Wrench. Pipe wrench.

Stilts. Adjustable extensions worn on the shoes of workers, such as plasterers or painters, to enable reaching higher than normal areas.

Stipple Finishing. Finish obtained by tapping surface with stipple brush before paint is dry.

Stippling. The method of shading by the making of separate marks or points.

Stipulated Sum Contract. A construction contract where the contract sum is a lump sum agreed amount; also called a Lump Sum Contract.

Stirrup. Reinforcing bars used in beams for shear reinforcement; typically bent into a U-shape or box-shape and placed perpendicular to the longitudinal steel.

Stirrup Pump. A portable hand pump stabilized with a foot bracket, used especially for fire-fighting.

Stirrup Tie. A stirrup that forms a complete loop, as differentiated from a U-stirrup, which has an open top.

Stock Dyeing. Dyeing raw fibers before they are carded (combed) or spun.

Stoke. A unit of kinematic viscosity, equal to the viscosity of a fluid in poises divided by the density of the fluid in grams per cubic centimeter.

Stockholder's Equity. Section of the balance sheet also known as Net Worth or Equity; representing the claim of the holders of common stock on the assets of the business.

Stone Facing Guard. Steel components applied to face of vertical stone construction to protect it from damage.

Stone. 1. Earthy or mineral matter of indeterminate size or shape such as rock. 2. Use of a carborundum stone to eliminate jagged and flaked edges of tile or masonry due to cutting

Stonemason. A mason who builds in stone.

Stonemasonry. Stonework; the work of a stonemason.

Stone Paver. Blocks of rock processed by shaping, cutting or sizing, used for driveways, patios and walkways.

Stoneware. A vitreous or semivitreous ceramic ware of fine texture, made primarily from nonrefractory fire clay.

Stonework. Masonry of stones and mortar.

Stool. A wood molding that provides an interior ledge at the sill of a window.

Stoop. A small unroofed porch.

Stop. A molding covering the crack between a door or window and its frame.

Stop Box or Curb Box. An adjustable cast iron box that is brought up to grade with a removable iron cover; by inserting a shutoff rod down into the stop box it is possible to turn off the water supply at the curb cock.

Stop, Door. See Door Stop.

Stop, Gravel. See Gravel Stop.

Stop Notice. A charge against construction funds in the hands of a property owner or a construction lender for the value of work or materials incorporated into a construction project; this is a lien on funds to be paid whereas a mechanic's lien is a lien on property; this is usually the only form of remedy to one who has done work for a Public Works Agency.

Storage Life. See Shelf Life.

Storage Tank Removal. The act or process of removing an old storage tank.

Storage Tank. A large receptacle used for the holding, transporting, or storage of liquids or gases.

Store. 1. A room or building for retail sales; a shop. 2. A reserve supply of goods.

Store Room. A room in which items are stored.

Storefront. 1. The facade which is constructed on the street side of a building or structure into which persons can enter and transact business. 2. A steel or aluminum tube frame and glass wall.

Storm Drain. A drain used for carrying rain water, cooling water, and sub-surface water, but not waste or sewage.

Storm Sash. An extra window usually placed on the outside of an existing window as additional protection against cold weather and air infiltration; also called a Storm Window.

Storm Sewer. A sewer used for conveying groundwater, rainwater, surface water, or similar nonpollutional wastes.

Storm Water Drainage System. The piping system used for conveying rainwater or other precipitation to the storm sewer or other place of disposal.

Storm Window. See Storm Sash.

Story. The space within a building between two adjacent floor levels or between a floor and a roof.

Story High. The height between floors or floor joists.

Story Pole. 1. A rod or pole that is the height of the story marked on it from joist to joist and also has the brick courses and any openings marked on it. 2. In ceramic tile installation, a long strip of wood marked at the appropriate joint intervals for the tile to be used; it is used to check the length, width, or height of the tilework; also called an Idiot Stick. 3. Wooden rod used to measure heights and locate rows of siding.

Straight Arch. See Jack Arch.

Straightedge. 1. A straight piece of lumber or metal that is used to rod mortar, align tile, or provide a straight or flush surface. 2. A flat wooden tool or rod, perfectly true, used to straighten the brown coat or screeds.

Straight-Grained Wood. Wood in which the fibers run parallel to the axis of the piece.

Straight Joint. The usual style of laying tile where all the joints are in alignment.

Straight-Line Depreciation. A method of calculating depreciation in which the same amount is depreciated each year.

Straight Peen Hammer. A hammer where the face opposite the flat face is shaped for cutting or grooving metal; the peen is flat and is perpendicular to the handle.

Straight Polarity. The arrangement of direct current arc-welding leads wherein the work is the positive pole and the electrode is the negative pole of the welding arc.

Straight Sheathing. Wood sheathing with the individual boards running at a 90 degree angle to the studs, joists, or rafters.

Strain. 1. Change in shape of a material due to stress. 2. To exert to the utmost. 3. To filter.

Strainer. Device such as a screen or filter used to retain solid particles while liquid passes through.

Strainer Basket. Device in skimmer and input side of pump used to catch large pieces of debris in pool water.

Strain Release. Movement along a fault plane; can be gradual or abrupt.

Strandboard. See Oriented Strand Board.

Stranded Conductor. A number of fine wires twisted around a center wire or core, used as a single electric conductor.

Stranded Wire. Fine wires twisted together in a group to create a larger stronger cable or wire.

Strap Channel. A U-shaped or L-shaped iron plate for connecting two or more timbers.

Strap Hinge. A surface hinge of which one or both leaves are of considerable length.

Strap, Malleable. See Malleable Strap.

Strap, Steel. See Steel Strap.

Strap Tie. A metal plate that fastens two parts together as a post, rod, or beam.

S-Trap. A plumbing fitting used to trap gases from a waste system.

Stratification of Air. Condition in which there is little or no air movement in room; air lies in temperature layers.

Strawberry. A small bubble or blister in the flood coating of a gravel-surfaced roof membrane.

Street. A public way to accommodate automobiles.

Street Ell. A 45 or 90 degree pipe elbow with male threads on one end and female threads on the other end; also called a Service Ell.

Street Removal. The act or process of removing old asphalt from a roadway.

Street Tree. A tree, planted in a specially prepared well, in the sidewalk or parkway adjacent to a public way like a street.

Strength. Capacity to resist force.

Strength Design. A design theory used for most reinforced concrete designs; a design in which members are designed to fail; safety is not provided by limiting stresses, but by using a factored load that is greater than the actual load.

Strength, Early. Concrete or mortar strength during the first 72 hours after placement.

Strength of Materials. A branch of mechanics and experimental physics dealing with stresses, strains, and the general behavior or materials and structural elements under the action of forces and moments.

Strength Ratio. In wood, the hypothetical ratio of the strength of a structural member to that which it would have if it contained no strength-reducing characteristics such as knots, slope-of-grain, shakes, or other defects.

Strength Reduction Factor. A factor introduced into the calculation of strength design in reinforced concrete, which increases actual loading, providing an added factor of safety.

Strength, Required. See Required Strength.

Stress. 1. The internal force of a body that resists external force. 2. The intensity of force per unit area.

Stress Diagram. A graphic representation of the value of the stresses in a structural member.

Stressed-Skin Panel. A panel consisting of two face sheets of wood or metal bonded to perpendicular spacer strips.

Stress-Grade Lumber. Structural lumber that has been graded and stamped with information to indicate the specific load it will support.

Stress Relief. See Perimeter Relief.

Stress-Relief Heat Treatment. Uniform heating of a structure or portion thereof to a sufficient temperature, below the critical range, to relieve the major portion of the residual stresses, followed by a slow uniform even cooling.

Stretch. The width of the area on which a painter will normally apply paint across a ceiling or down a side wall.

Stretcher. A brick that is laid on its flat side and shows its face on the surface of the wall.

Stretcher, Hospital. See Hospital Stretcher.

Stria. A striped surface effect obtained by loosely twisting two strands of one shade of yarn with one strand of a lighter or darker shade.

Strict Liability. Liability without fault, imposed by the courts for social reasons to distribute the risk of injury to consumers by imposing on the manufacturers of products liability for injury caused by the use of those products even though the manufacturer is not shown to be at fault.

Strike. 1. An organized refusal by employees to work until some grievance is remedied. 2. A metal plate or box that is pierced or recessed to receive the bolt or latch when projected; also called the Keeper or Strike Plate.

Strike Plate. See Strike, 2.

Strike Slip Fault. A geological fault which runs parallel to the strike of associated strata.

Striking In. Materials used in finishing are said to strike in when they soften undercoats and sink into them.

Striking Joints. A process of removing excess grout from the joints by wiping with a sponge or cloth or scraping, compacting or rubbing with a curved instrument.

String. 1. A light cord or twine. 2. See Stringer

String Course. A horizontal course of brick or stone on the face of a building; when continuous with a row of window sills or lintels, is referred to as a Sill Course or a Lintel Course; also called a Belt Course.

Stringer. 1. The sloping wood or steel member supporting the treads of a stair. 2. A long horizontal timber in a structure supporting a floor.

Stringer Bead. A type of weld bead made without appreciable transverse oscillation.

Stringing Mortar. The procedure of spreading enough mortar on a bed to lay several masonry units.

String Level. A small spirit level attached to a string line for establishing level lines on a construction site.

Stringline. A string used to establish a line or to mark a line on a construction site.

String Wire. Wire used on open stud construction, placed horizontally around building to support weather-proofing paper.

Strip. 1. Complete removal of an old finish with paint removers. 2. The removal of wallpaper. 3. To remove forms from set concrete surfaces. 4. To damage the threads of a bolt or screw. 5. A long, narrow, piece of a material.

Strip, Cant. See Cant Strip.

Strip, Chamfer. See Chamfer Strip.

Strip Flooring. Hardwood flooring laid in strips, usually tongued and grooved.

Strip Footing. See Continuous Footing.

Striping. Edge painting prior to priming.

Strip Lath. A narrow strip of diamond mesh metal lath sometimes applied as reinforcement over joints between sheets of gypsum lath, at the juncture of two different base materials, and at corners of openings.

Strip Mopping. The mopping of roofing bitumen in parallel bands, generally about 8 inches in width with 4 inch unmopped spaces.

Stripping. 1. The technique of sealing the joint between built-up roofing membrane and metal using one or two plies of felt or fabric and hot or cold applied bitumen.2. The method of taping the joints between insulation board.

Strips. Bands of reinforcing bars in flat slab or flat plate construction.

Stroboscope. An instrument used for determining speeds of rotation by shining a flashing bright light at measured intervals so that a rotating object appears stationary.

Stroke. A single pass with a spray gun in one direction.

Struck Joint. In masonry, a mortar joint which is formed with a recess at the bottom of the joint; it is the reverse of the weathered joint; this joint is used extensively, but chiefly for interior wall surfaces, since it is inferior for outside joints, because of its lack of weather-resisting qualities; the recess at the bottom allows water from rain or snow to seep into the wall.

Structural Bond. The interlocking pattern of masonry units used to tie two or more wythes together in a wall.

Structural Clay. A term applied to various sizes and kinds of hollow and practically solid building units, molded from surface clay, shale, fire clay or a mixture of these materials and laid by masons.

Structural Component. A member of a structural system.

Structural Composite Lumber. Lumber in small pieces glued together to make structurally sound larger members, as in glued laminated beams or columns.

Structural Defects. 1. Defects in a structure that pertain to the structural elements which may cause an unsafe condition. 2. In tilework, cracks or laminations in the body of the tile which detract from the aesthetic appearance or the structural soundness of the tile installation.

Structural Design. The design of the structural system, analysis of support of all dead and live loads, and calculation of sizes and arrangement of all members.

Structural Diaphragms. Structural members, such as floor and roof slabs, which transmit inertial forces to lateral-force-resisting members.

Structural Engineer. A civil engineer who specializes in the design and analysis of structures.

Structural Fee. The amount of money charged for structural engineering work.

Structural Frame. The columns and girders, trusses, beams and spandrels which directly connect to the columns and all other members necessary to the stability of the building; secondary members are those members of the floor or roof panels which are not connected to the columns and are not part of the structural frame.

Structural Glazed Tile. A hollow clay masonry unit with glazed faces.

Structural Lightweight Concrete. See Concrete, Structural Lightweight.

Structural Lumber. 1. Wood members of a structural system which are manufactured by sawing, resawing, passing lengthwise through standard planing machine, crosscutting to length, but without further manufacturing. 2. Lumber that is 2 or more inches thick and 4 or more inches wide; intended for use where strength is required; the grading of structural lumber is based on the strength of the piece and its use.

Structural Lumber Exposed To View. Structural lumber which is exposed to view in finished spaces or at the exterior of the building.

Structural Pipe. Pipe used in a structure to transfer imposed loads to the ground.

Structural Plywood Diaphragm. Structural plywood sheathing on floors, roofs, or walls which provides shear strength to resist wind and earthquake loads.

Structural Plywood. The highest grade of plywood; plywood of exterior structural grade, secured to top side of floor joists used to create rigidity in building superstructure and also to provide a smooth and even surface to receive finish floor covering.

Structural Shear Wall. Wall constructed of studs and structural-use panel sheathing that in its own plane resists shear forces resulting from applied wind, earthquake, or other transverse loads or provides frame stability.

Structural Steel. Steel hot-rolled into variety of shapes for use as load-bearing structural members.

Structural Steel Door Frame. Door frame fabricated of structural steel shapes, such as channel.

Structural Steel, Carbon Grade. Steel having either (1) no specified minimum content of alloying elements; (2) specified minimum copper content not exceeding 0.40 percent, or (3) maximum specified percentage contents as follows. manganese 1.65, silicon 0.60, copper 0.60.

Structural Steel, Corrosion Resistant. Steel having chemical composition specifically developed to impart higher mechanical property values and with resistance to atmospheric corrosion than is four times greater than that of conventional carbon structural steel; this steel weathers (oxidizes).

Structural Steel, High-Strength. Steel having chemical composition specifically developed to impart higher mechanical property values than is obtainable from conventional carbon structural steel.

Structural Steel, Quenched and Tempered. Steel which is quenched (rapidly cooled) and tempered (reheated, quench hardened, or normalized ferrous alloy to temperature below transformation range, followed by cooling of alloy at rate desired); primarily used for welded bridge construction.

Structural Steel Tubing. Hollow structural tubing, usually circular, square, or rectangular.

Structural Subflooring. Flat sheet material of structural use panel secured to top side of floor joists used to create rigidity in building superstructure and serve as base to receive flooring.

Structural System. All of the supporting parts of a building taken as an integrated whole.

Structural Timbers. Pieces of wood of relatively large size, with a cross section greater than 4 inches x 6 inches, the strength of which is the controlling element in their selection and use.

Structural Trusses. Assemblages of wood, steel, or reinforced concrete members subjected primarily to axial forces.

Structural Tube. A hollow metal product used to carry imposed loads in a building or structure.

Structural Tubing Railing. A metal railing made of structural metal tubing.

Structural Underlayment. Flat sheet material of structural use panel secured to top side of floor joists used to create rigidity in building structures and to provide smooth and even surface to receive finish floor covering.

Structural Use Panel. Flat sheet material of either plywood, oriented strandboard, waferboard, or particleboard.

Structural Use Panel Roof Sheathing. Flat sheet material of structural-use panels secured to exterior side of roof rafters or trusses, used to create rigidity in building superstructure and serve as base to receive roofing.

Structural Wall Sheathing. Flat sheet material of structural-use panel secured to exterior side of exterior wall studs, used to create rigidity in building superstructure and serve as base to receive siding or veneer construction.

Structural Wood Member. Member of a wood structural system.

Structure. That which gives form to something and works to resist changes in the form due to the action of various forces; a building.

Structure Borne Sound. A condition in which the sound waves are carried by a solid material.

Strut. 1. A compression member; a column; it can be placed at any angle, not just vertically. 2. An element of a structural diaphragm, used to provide continuity around an opening in the diaphragm.

Stub Out. A capped pipe or conduit provided for future connection or extension of water, gas, electrical, or other utility.

Stub Tie. A method of attaching metal lath to furring channels, where the wire is twisted and cut off at the twist.

Stub Truss. A truss which has one end truncated by being cut vertically.

Stucco. 1. A cement plaster used for coating exterior walls and other exterior surfaces of buildings. 2. A plaster used for interior decoration and finish work. 3. A manufacturing term for calcined gypsum prior to addition of water.

Stucco Edge. In gypsumboard, a break in the fold or lapped seal between the face and back papers during the manufacturing process which allows the wet slurry to ooze onto edges and face.

Stucco Lath. Wood or metal lath strips which form the base for the application of a cement plaster on an exterior wall surface.

Stucco Mold. A wood or metal molding to retain the edges of stucco and provide a finish as at doors and windows.

Stucco Netting. Galvanized woven fabric or chicken wire used in the lathing process for exterior stucco over light wood framing construction.

Stucco Netting. See Woven Wire Fabric.

Stud. 1. A vertical wall framing member, wood or steel, may be load bearing or non load bearing. 2. A vertical member to support sheathing of concrete forms. 3. A headed steel device used to anchor steel plates or shapes to concrete members.

Studding. The framework of a partition or wall of a house; usually referred to as 2 by 4's.

Studio Apartment. A small living unit consisting of a combination living-dining-kitchen-sleeping room and a bathroom

Stud Removal. The act or process of demolishing and carrying away old stud walls from a building or structure.

Stud Shear. A hand tool for cutting metal studs.

Stud, Screwable Type. See Screwable Type Stud.

Stuffer Warp. Yarn which runs lengthwise in the carpet but does not intertwine with any filling yarns; serves to give weight, thickness and stability to the fabric; also called Stuffer Yarn.

Stuffer Yarn. See Stuffer Warp.

Style. A distinctive manner in respect to appearance, as in the Greek style or the modern style.

Stylobate. A continuous rectangular base on which a row of architectural columns are placed.

Styrene-Butadiene Resin. Synthetic rubber resin; liquid styrene and butadiene gas, copolymerized to form chemical-resistant product with excellent film-forming properties.

Styrene-Rubber (SR) Pipe & Fittings. Plastics containing at least 50% styrene compounding materials but not more than 15 percent acrylonitrile.

Subbase Course. The course in the asphalt pavement structure immediately below the base course; if the subgrade soil is of adequate quality, it may serve as the subbase.

Sub-bidder. A subcontractor who submits a bid to a prime contractor.

Subcontractor. 1. A contractor whose contract is with the general contractor, not the owner. 2. An individual or entity contracting to perform part or all of another's contract.

Subcontractor List. A list of the subcontractors a general contractor contemplates using, submitted for the owner's approval, as required by some construction contracts.

Subcooling. Cooling of liquid refrigerant below its condensing temperature.

Subdiaphragm. A portion of a larger wood diaphragm designed to anchor and transfer local forces to primary diaphragm struts and the main diaphragm.

Subdivider. One who divides a tract of land into building lots, streets, and open spaces.

Subdivision. The division of a tract of land into building lots, streets, and open spaces.

Subfloor. A wood floor which is laid over the floor joists and on which the finished floor is laid.

Subflooring. Certain material, like plywood, that is installed on the floor joists of a building or structure, onto which the walls and finished flooring is attached.

Subgrade. The soil prepared to support a structure or a pavement system; it is the foundation for the pavement structure; also called Basement Soil or Foundation Soil.

Subgrade, Improved. Subgrade, improved as a working platform (1) by the incorporation of granular materials or stabilizers such as asphalt, lime, or portland cement, prepared to support a structure or a pavement system, or (2) any course or courses of select or improved material placed on the subgrade soil below the pavement structure; subgrade improvement does not affect the design thickness of the pavement structure.

Subjacent Support. The right of land to be supported by the land adjacent to or below it.

Subject Property. The property that is the subject of a real estate appraisal report.

Sublease. Lease of property by a tenant to a subtenant; also called Subletting.

Subletting. See Sublease.

Sublimation. Conversion of a solid to a gas by heating and from a gas to a solid by cooling, for example, dry ice (carbon dioxide).

Submarginal. Falling below the accepted or required standard of quality.

Submarine. Of or below water.

Submersible Pump. A centrifugal pump, usually driven by electricity or compressed air, that will operate when submerged.

Submission Agreement. An agreement to submit an existing dispute to arbitration.

Submittals. Various documents submitted by the contractor to the owner or architect, such as shop drawings, material samples, subcontractor list, and insurance certificates.

Submittal Schedule. A time schedule prepared by a contractor showing when each of the specified submittals will be sent to the architect or owner.

Subordination Clause. A provision in a mortgage or trust deed that acknowledges that it is subordinate to another loan or will allow a future loan to be a superior lien on the property.

Subpoena. A legal order requiring a person to appear in a court, arbitration hearing, or deposition to give testimony.

Subpoena Duces Tecum. A subpoena requiring the production of specified documents.

Subpurlin. A small roof framing member that spans between joists and purlins.

Subrogation. The assumption by a third party, such as an insurance company, of another's legal right to collect a debt or damages.

Subsidiary Ledger. A supporting ledger consisting of a group of similar accounts, the total of which is in agreement with a controlling account in the general ledger.

Subsoil Drain. A drain used for collecting sub-surface or seepage water and carrying the water to where it can be disposed.

Substance. Any form of matter or material.

Substantial Compliance. A doctrine that may excuse minor violations of licensing statutes.

Substantial Performance. The doctrine that a contract is enforceable by a party who has not strictly complied with all requirements of a contract, but who has substantially complied therewith.

Substitutions. A substitution of materials for those originally specified for the project.

Substrata. Substrate.

Substrate. 1. A lower level, upon which other materials are applied, such as paint, flooring, or roofing. 2. The underlying support for ceramic tile or other finish-type installation. 3. The base or concealed layer of gypsum board in a composite assembly.

Subsubcontractor. An individual or entity contracting to perform part or all of a subcontractor's contract.

Subsurface Investigation. See Geotechnical Investigation.

Subsurface Right. The right to use the subterranean part of a parcel of land, such as for tunneling, extraction of minerals, and pipe lines, usually given in the form of an easement.

Subterranean. Below the surface of the earth.

Subterranean Termite. See Termite.

Subway. 1. An underground railway. 2. A passage under a street for pedestrians or utility lines. 3. An underpass.

Suction. The power of absorption possessed by a plastered surface.

Suction Diffuser. A centrifugal pump which forces water out of an area or surface.

Suction Feed. 1. In a sandblast gun, one in which the abrasive is siphoned to the nozzle. 2. In a spray gun, one in which the fluid is siphoned to the spray head.

Suction Line. Tube or pipe used to carry refrigerant gas from evaporator to compressor.

Suction Lines. In a swimming pool, pipework supplying the main circulating pump.

Suction Piping. That portion of the circulation piping located between the pool structure and the inlet side of the pump and usually including the main outlet piping, skimmer piping, vacuum piping, and surge tank piping.

Suction Pressure. Pressure in low-pressure side of a refrigerating system.

Suction Pressure Control Valve. Device located in the suction line which maintains constant pressure in an evaporator during the running portion of a cycle.

Suction Service Valve. Two-way manually operated valve located at the inlet to compressor; it controls suction gas flow and is used to service unit.

Suction Side. Low-pressure side of the system extending from the refrigerant control through the evaporator to the inlet valve of the compressor.

Suction Spotting. Spotting of paint job caused by oil in new coat being absorbed by spots or porous areas of surface.

Suction Valve. Valve in refrigeration compressor which allows vaporized refrigerant to enter cylinder from suction line and prevents its return.

Sue and Labor Clause. A clause in a property insurance policy that authorizes the insured to expend funds to protect the insured property.

Sulfate Attack. Deleterious chemical or physical reaction between sulfates in ground water or soil and certain constituents in cement, which result in expansion and disruption of the concrete.

Sulfate Resistance. Ability of cement paste and aggregate to withstand sulfate attack.

Sulphuric Acid. A dense corrosive oily colorless fluid that is a vigorous oxidizing and dehydrating agent; also called Oil of Vitriol.

Summary Judgment. A judgment awarded on the basis of affidavits and legal briefing rather than on the basis of evidence introduced at a trial.

Summer. 1. A horizontal beam such as a lintel. 2. A stone cap on a pier.

Sump. A sump is a specially made receiving tank to receive wastes or sewage by gravity; from the sump the wastes or sewage is lifted by pump or ejectors to be discharged into the building drain or building sewer.

Sump Pot. See Main Outlet.

Sump Pump. A small capacity pump that empties pits receiving groundwater, sewage, or liquid waste.

Sundays. Place skipped when applying finishing materials to a surface.

Sundial. An instrument showing the time by the shadow of a pointer (called a gnomon or stylus) cast by the sun on a graduated disc.

Sunshade. A device that protects from the sun's rays, such as an awning or eyebrow.

Superchlorination. Heavy dose of chlorine added to pool water to burn out nitrogen compounds when bacteria, algae, or ammonia build-up cannot be reduced by normal treatment.

Superconductivity. The property of zero electrical resistance in some substances at very low absolute temperatures.

Superelevation. The additional elevation of the outer side of a curved section of a roadway to overcome centrifugal forces of vehicles.

Superheat. 1. Temperature of vapor above its boiling temperature as a liquid at that pressure. 2. The difference between the temperature at the evaporator outlet and the lower temperature of the refrigerant evaporating in the evaporator.

Superheater. Heat exchanger arranged to take heat from liquid going to evaporator and using it to superheat vapor leaving evaporator.

Superimposed Load. Loads and stresses added to the dead load of the structure; live load.

Superintendent. 1. An individual who is at the top level of a construction team in the field. 2. A person with executive oversight, often of a board or building complex.

Superstructure. 1. The part of a building above the foundation. 2. In bridge construction, the concrete deck or traffic surface of the bridge.

Supervision. A contractor's scheduling, watching, and directing of workers during the execution of work.

Supplementary Conditions. A part of the contract documents that may supplement, delete, add to, or amend the general conditions.

Supplementary Drawings. Additional drawings which augment the contract drawings.

Supplier. One who supplies construction materials to a project.

Supply Air Duct. The duct from a heating, cooling, or ventilation system carrying the treated air to the space of use.

Supply Tank. Separate tank connected directly or by a pump to the oil-burning appliance.

Supports. 1. The structural foundation for essential building elements. 2. Devices for supporting and securing pipe, fixtures, and equipment.

Surcharge. 1. An increase in the lateral earth pressure of a retaining wall, caused by a vertical load behind the wall. 2. A load placed over an area to compact it or change its characteristics.

Surety. One who undertakes to guarantee performance by another; usually an insurance company.

Surety Bond. A bond guaranteeing the performance of a contract or an obligation; see Performance Bond, Labor and Material Bond, and Completion Bond.

Surface. The outside layer of a material or body.

Surface Bolt. A door bolt that is mounted on the surface of a door with the bolt sliding into a keeper in the top or side of the door frame or into the sill or floor.

Surface Depression. A mark or indentation on the gypsum board surface.

Surface Drying. Drying of a finishing material on top while the bottom remains more or less soft.

Surfacer. A paint used to smooth the surface before finish coats are applied.

Surface Hinge. One having both leaves applied to the surface of the door.

Surface Lumber. Lumber that is dressed by running it through a planer.

Surface Metal Raceway. Similar to a wireway, used for electrical conductors and communication wiring.

Surface Moisture. Free moisture retained on the surfaces of aggregate particles which becomes part of the mixing water in the concrete mix.

Surface Plate. Tool with a very accurate flat surface.

Surface Preparation. All operations necessary to prepare a surface to receive a coating of paint.

Surface Rights. The right to use the surface of real property, usually in the form of an easement.

Surface Skimmer. In a swimming pool, a device designed to continuously remove surface film and water and return it through the filter as part of the recirculation system, usually incorporating a self-adjusting weir, a collection tank, and a means to prevent air lock of the pump; also called a Mechanical Skimmer, an Automatic Skimmer, or a Recirculating Overflow.

Surface Tension. The tension of the surface film (meniscus) of a liquid, tending to minimize its surface area.

Surface Water. Water carried by an aggregate except that held by absorption within the aggregate particles themselves; water in addition to SSD water.

Surface Waters. Rain water collected and running on the surface of the land rather than being confined to drains and water courses.

Surficial. Relating to a surface.

Surfacing. In welding, the deposition of filler metal on a metal surface to obtain desired properties or dimensions.

Surfacing, ACM. Asbestos-containing material that is sprayed-on, troweled-on or otherwise applied to surfaces, such as acoustical plaster on ceilings and fireproofing materials on structural members, or other materials on surfaces for acoustical, fireproofing, or other purposes.

Surge. 1. To rise or fall rapidly, as electrical current. 2. Regulating action of temperature or pressure before it reaches its final value or setting.

Surge Arrestor. A electrical device that minimizes or eliminates the deleterious effect of a sudden rise in voltage or current; also called a Surge Suppressor.

Surge Chamber. In airless spray equipment, a device to eliminate uneven fluid flow.

Surge Suppressor. Surge Arrestor.

Surge Tank. Container connected to the low-pressure side of a refrigerating system which increases gas volume and reduces rate of pressure change.

Surgical Chronometer. An instrument used to measure time with great accuracy, used in medical applications.

Surgical Light. A light source used by surgeons for its illumination abilities for close work.

Surgical Scrub. Sterilized clothes worn by physicians in surgery.

Surgical Station. The area in a hospital where surgeons operate and the location of necessary implements for surgeries.

Surgical Table. A surface where surgery is performed.

Surplus. The earnings of a business from previous years which were allowed to accumulate in the business.

Surround. An area or substance that surrounds something, such as the frame around a wall opening.

Surveillance System. An interacting system of devices, such as cameras, that detect presence in and often videotape an area.

Survey. 1. To examine some topic or material to determine its condition, situation, or value. 2. The precise measurement of a parcel of land to determine all its dimensions, both vertical and horizontal.

Surveying. A branch of applied mathematics, including geometry and trigonometry, that aids in determining and delineating the form, extent and position of land.

Suspended Ceiling. A finish ceiling that is hung on wires from the structure above.

Suspended Ceiling Removal. The act or process of removing the modular units and skeleton frame of an old suspended ceiling.

Suspended Structure. A structure supported principally by tension members or carrying its loads principally in tension.

Suspended Unit Heater. A unit heater that is hung from the roof framing of an open space like a factory or workshop.

Suspension Bridge. A bridge with roadway suspended from cables supported by structures at each end.

Suspension of Work. The temporary stopping of the work, a right reserved by owners in some construction contracts.

Suspension System. A system, usually incorporating heavy gage wire, designed to support the finished ceiling membrane in place at a pre-determined distance below the structural framing.

Sustainability. 1. Capable of supporting the weight for a long period. 2. Ability to keep up, prolong, keep going for a long time.

Sustainable Architecture. Architectural design based on the principles of sustainable development.

Sustainable Building. Construction that is based on energy conservation and reduction in use of non-renewable resources.

Sustainable Development. Construction utilizing materials that are replaceable, like wood, and minimizing use of non-replaceable materials and energy resources.

SW. Severe Weather; see Grade SW Brick.

Swage. 1. A tool used for metal forming by holding it on the work or the work on it and striking with a hammer or sledge. 2. To increase or decrease the diameter of a pipe by using a special tool which is forced around the pipe.

Swaging. Enlarging one tube end so end of other tube of same size will fit within.

Swale. A low-lying, often depressed and swampy, area of land; an open ended swale can be created for use as a land drainage device.

Swale Excavation. The digging up of low-lying land such as a small meadow, swamp, or marshy depression.

Swamp Cooler. Evaporative type cooler in which air is drawn through porous mats soaked with water.

Swash Plate-Wobble Plate. Device used to change rotary motion to reciprocating motion; used in some refrigeration compressors.

S-Wave. Shear wave, produced essentially by the shearing or tearing motions of earthquakes at right angles to the direction of wave propagation; see Seismic Wave.

Sway Brace. Diagonal bracing to prevent lateral movement caused by horizontal forces.

Sweat Out. A defective condition occasionally occurring in gypsum plaster; characterized by a soft, damp area remaining after the surrounding area has set hard. Often caused by insufficient ventilation which inhibits normal drying.

Sweat Soldering. Method of soldering in which the parts to be joined are first coated with a thin layer of solder (tinned) then joined while exposed to heat.

Sweating. 1. Condensation of moisture from air on cold surface. 2. Method of soldering in which the parts to be joined are first coated with a thin layer of solder; see Tinning.

Swedish Putty. See Spackling Compound.

Sweet Water. Tap water.

SWI. Steel Window Institute.

Swimming Load. Number of people using pool at a given time.

Swimming Pool. A tank of concrete, steel, or plastic suitable for swimming.

Swimming Pool, Private. All constructed or prefabricated pools which are used as a swimming pool in connection with a single family residence, and available only to the family of the householder and their private guests.

Swimming Pool, Public. Any constructed or prefabricated pool other than a private swimming pool.

Swing Bridge. A moveable bridge, being pivoted in the middle to move out of the way of river traffic.

Swing Gate. The operable member of a fence system that is hinged for opening and closing.

Swing Joint. A joint in a threaded pipe line which permits motion in the line of a plane normal to the direction of one part of the line.

Swirl Texturing. A method of applying gypsumboard texturing material in a decorative circular pattern.

Switch. A device used to continue, disrupt, or redirect an electrical current.

Switchboard. A large dead-front, single panel, frame, or assembly of panels, designed and manufactured as a unit, and which may house bussings, externally operable fused switches, circuit breakers, or other protective or regulating devices and associated instrumentation.

Switch, Dimmer. See Dimmer Switch.

Switch, Fusible. See Fusible Switch.

Switchgear. A freestanding assembly including primary (disconnect) switches, secondary (feeder) switches, and overcurrent protection device (fuses and circuit breakers).

Switch, Oil. See Oil Switch.

Switch Plate. A cover plate covering an electric switch.

Switch, Safety. See Safety Switch.

Switch, Transfer. See Transfer Switch.

Swivel. A device joining two parts so that the moving part can pivot freely.

Sylphon Seal. Corrugated metal tubing used to hold seal ring and provide leakproof connection between seal ring and compressor body or shaft.

Symbol. A mark, sign, or character taken as the conventional representation of some object, idea, function, or process; for example, the letters standing for the chemical elements or mathematical constants.

Symmetrical. Having identical parts on both sides of a center line.

Syndicate. A combination of individuals or commercial firms to promote some common interest.

Synergy. A combined effect or action.

Synthesis. The process of combining a set of component elements into a whole.

Synthetic Fiber. A fiber made by a chemical process. for example, nylon, polyester, dacron, rayon, acetate, and acrylic.

Synthetic Resin. A solid or liquid organic compound made by polymerization, the material of which a plastic is made, including epoxy resin, acrylic resin, and many other polymers; chemical compounds react to form synthetic resins.

Synthetic Rubber. Made by polymerizing hydrocarbons; see Neoprene.

Synthetic Wall Mud. An alternative to conventional plaster mortar, contains a lightweight aggregate in place of sand; available in 50 pound bags.

System. An ordered assemblage.

Systematic Errors. Cumulative errors.

T

T. Transmittance.

T&G Siding. Tongue and groove exterior siding.

T&G. 1. Tongue and Groove. 2. Tongued and grooved.

T&M. Time and Material.

Taber Abrader. An instrument used to test the abrasion resistance of a material.

Table, Laboratory. See Laboratory Table

Table Saw. A power saw in which the saw remains stationary on a table and the material to be cut is passed under it.

Tableware. All utensils and decorative articles used on the table for meal service.

Tachometer. An instrument for measuring the speed of a rotating shaft or a moving vehicle.

Tack. 1. Short sharp pointed nail with large head used to secure thin or woven materials to wood and similar materials. 2. Degree of stickiness.

Tack Board. A board, often of cork, for tacking up notices and displays.

Tackboard. Soft panel used for attachment of items with thumbtacks.

Tack Coat. Application of asphaltic material on old surface to insure its bond to new construction.

Tack Hammer. A small hammer for installing tacks.

Tackiness. 1. Stickiness. 2. When a painting material dries out, gels, or sets up, it loses tackiness or stickiness.

Tack Rag. A piece of cheesecloth or cotton rag moistened with thinned varnish, used by painters to pick up small particles of abrasive dust.

Tack Strip. A wooden strip containing a series of inclined barbs for grasping the edges when installing carpeting.

Tack Weld. A small weld to hold steel pieces together temporarily.

Tack Welding. Before welding, pipes are properly aligned and then tack welded; special line-up clamps are used to insure correct alignment.

Tacky. 1. Not quite dry; sticky. 2. Shabby; seedy.

Tail Line. Short piece of blast hose smaller than the main hose to permit better maneuverability.

Tail Pipe. Outlet pipe from the evaporator.

Tails. In airless spray painting, finger-like spray pattern.

Take-Off Man. Someone who can read blueprints and is familiar with the specifications; makes notes of special details concerning the project after gathering the necessary information and then estimates the quantities of labor, materials, equipment and special items needed to complete the job; also called a Take-Off Person or a Quantity Surveyor.

Take-Off Person. See Take-Off Man.

Takeout Loan. See Permanent Loan.

Talc. 1. Any crystalline form of magnesium silicate that occurs in soft flat plates, used as a lubricator or dryer. also called Talcum

Powder or Soapstone. 2. A fine powder applied to the back of gypsumboard to minimize friction and permit easier sliding of bundles. 3. A hydrous magnesium aluminum silicate used as an extender in paints; helps make paint smooth.

Talcum Powder. See Talc.

Tall Oil. A blend of resin and oil acids obtained as a by- product from the sulfate process for making paper.

Tamp. Pound or press soil to compact it.

Tamper. A tool for compacting soil in spots not accessible to rollers.

Tamping Rod. Used in preparing concrete cylinders for testing, a round, smooth, straight steel rod, 5/8 inch in diameter and approximately 24 inches in length, having the tamping end rounded into a hemispherical tip, the diameter of which is 5/8 inch.

Tandem. A device with two parts, one behind the other.

Tangent. 1. A straight line that touches a curve at one point but continues on without crossing it. 2. The trigonometric ratio of the sides opposite and adjacent to an angle in a right-angled triangle.

Tangential. 1. Acting along or lying in a tangent. 2. Strictly, coincident with a tangent at the circumference of a tree or log, or parallel to such a tangent; in practice, however, it often means roughly coincident with a growth ring; a tangential section is a longitudinal section through a tree or limb perpendicular to a radius; flat-grained lumber is sawed tangentially.

Tangible. Perceptible by touch; palpable; definite; clearly intelligible.

Tank, Supply. See Supply Tank.

Tank White. Good hiding, self-cleaning white paint for exterior metal converging lines.

Tap-a-Line. See Saddle Valve.

Tap Box. The electrical box where the public service electrical supply line is connected with a branch to serve a particular building or structure.

Tape. 1. A tape measure, marked off in measuring units. 2. A narrow woven fabric. 3. A strip of narrow transparent or opaque material coated with adhesive for fastening, sticking, or insulating. 4. Gypsumboard joint reinforcing tape. 5. A plastic reinforcing mesh or paper used to reinforce angles and to bridge lath joints in veneer plastering.

Tape Creaser. In gypsumboard joint finishing, a hand tool used in folding joint tape for use in inside corners.

Taper. 1. A gradual and uniform decrease in size, as of a round or rectangular piece or hole. 2. The factory edge that is progressively reduced in caliper from the face to the outer edge allowing for the concealment of joint tape below the plane of the gypsum board surface.

Tape Recorder. A mechanical device which makes sound recordings on magnetic tape.

Tapered Edge. See Taper.

Tapered Edge Strip. 1. A tapered insulation strip used to elevate the roof at the perimeter and at curbs that extend through the roof. 2. A tapered insulation strip used to provide a gradual transition from one thickness of insulation to another.

Tapered Pattern. Elliptical shaped spray pattern; a spray pattern with converging lines.

Tapered Steel Girder. A fabricated steel plate girder with a sloping top flange.

Tapestry. A looped pile fabric woven on the velvet loom.

Tape Test. A particular type of adhesion test.

Taping Compound. Gypsumboard joint compound especially formulated to embed joint tape.

Taping Tools. See Ames Tools.

Tapping. 1. The process of forming internal screw threads in a drilled hole. 2. The tapping or pounding of a brick or other masonry unit down into the bed of mortar.

Tapping Tile. An inspection technique whereby a coin, key, or other small metallic object is tapped against an installed tile to determine by sound whether the tile is completely bonded to its backing; a hollow sound occurs if the tile is not set properly; tilesetters often tap the tile with a pointing trowel to determine that a good bond has been achieved.

Tapping Valve. A device to open or close a duct, pipe, or other passage, or to regulate flow; it is inserted into an existing pipeline by piercing the wall of the pipe and thus tapping into the flow.

Taproot. A large root that grows downward from the base of a tree.

Tar. A dark thick inflammable liquid distilled from wood or coal and used as a sealant, a wood preservative, and for roofing and road paving.

Tare. The weight of an empty truck or container.

Tarp. See Tarpaulin.

Tarpaulin. A waterproofed canvas or other material used for protecting construction projects, athletic fields, goods, or other exposed objects or areas; also called a Tarp.

Tarred Felt. A felt that has been saturated with refined coal tar.

Taupe. A brownish gray.

Tax. A charge imposed by authority upon persons or property for public purposes.

Tax Shelter. The offsetting of investment income by non-cash charges, such as reserves for depreciation, for the purpose of reducing income tax.

T-Bar Ceiling System. A suspended ceiling system where the members are T-shaped metal that will accept acoustical board panels, lighting fixtures, or air conditioning registers and grilles.

T-Beam. A reinforced concrete beam that contains of a portion of the slab above and which the two act together.

TCA. Tile Council of America Research Center.

Teak. Yellowish-brown wood from the teak tree; used primarily for shipbuilding, furniture, and wood finish for buildings.

Teak Veneer. Thin sheets of teak, a dark wood, used for plywood or other finishes.

Tear-Off. A roofing repair job entailing complete removal of the old roof and insulation down to the deck.

Tear Strip. Heavy paper ribbon under bundling tape to facilitate opening of bundles.

Technically Infeasible. With respect to an alteration of a building or a facility, that it has little likelihood of being accomplished because existing structural conditions would require removing or altering a load-bearing member which is an essential part of the structural frame; or because other existing physical or site constraints prohibit modification or addition of elements, spaces, or features which are in full and strict compliance with the minimum requirements for new construction and which are necessary to provide accessibility.

Technical Competence. Knowledge of business and trade practices to successfully accomplish purposes of business.

Technobabble. Incomprehensible technical jargon.

Technology. 1. Applied science. 2. Technical language.

Tedlar. A registered trademark name of du Pont Co.; an elastomeric membrane consisting of polyvinyl fluoride

Tee. A metal or precast concrete member with a cross section resembling a T.

Tee, Bulb. Rolled steel in the form of a T with a formed bulb at the end of the web.

Tee Joint. A joint between two members located approximately at right angles to each other in the form of a T.

Tee Member. A metal or precast concrete member with a cross section resembling a T.

Tee Square. A 48-inch long T shaped metal guide used in scoring gypsumboard to job size requirements; also called a Hanger's Tee.

Tee Turn. Turn Piece.

Tee Weld. Weld in a joint between two members located approximately at right angles to each other in the form of a T.

Telamon. A male figure used as a supporting column.

Telescopic. Having parts that slide or pass within one another.

Telescoping Bleacher. A seating structure, usually in a gymnasium, that folds back into itself.

Teller Window. The opening in a wall partition through which transactions in a financial institution take place.

Temblor. Earthquake.

Temper. 1.The condition of metal as regards hardness and elasticity. 2. To mix plaster to a workable consistency. 3. To mix, by the adding of water to make the mortar the right consistency for use.

Tempera. 1. A process of fine art painting in which an albuminous or colloidal medium, as egg yolk, is employed as a vehicle instead of oil; poster color; water color. 2. A water-thinned or water-emulsion paint.

Temperature. 1. Degree of heat in a body as measured by a thermometer. 2. Measurement of speed of motion of molecules.

Temperature Bars. See Temperature Reinforcement.

Temperature Control. Temperature-operated thermostatic device which automatically opens or closes a circuit.

Temperature, Curing. See Curing Temperature.

Temperature-Humidity Index. Actual temperature and humidity of air sample compared to air at standard conditions.

Temperature Inversion. The phenomenon of a layer of warmer air trapping a layer of cooler air closer to the ground thus restricting its rise and with it trapping any air pollutants.

Temperature Reinforcement. Steel reinforcing bars distributed throughout the concrete to minimize cracks due to temperature changes and concrete shrinkage, also called Temperature Bars or Temperature Steel.

Temperature Rise. The increase of concrete temperature caused by heat of hydration and heat from other sources.

Temperature Steel. See Temperature Reinforcing.

Tempered Air. See Conditioned Air.

Tempered Glass. Glass that has been treated to increase it toughness and its resistance to breakage.

Template Hardware. Items of hardware that are made to template as to spacing of all holes and dimensions, as in hinges, locks, exit devices, and closers for hollow metal doors and frames.

Template. A gauge, pattern, or mold used as a guide to the form of a piece being made.

Temple. A building devoted to worship.

Temporary Centering. Temporary forming used during the construction process for support of a masonry arch or a concrete slab.

Temporary Closure. Temporary construction used during the construction process to close openings to provide weather protection or to separate construction areas from occupied areas.

Temporary Facility. A structure erected for temporary use.

Temporary Restraining Order. An order of brief duration issued by a court pending a hearing on the merits.

Temporary Services. The utilities (water, electricity, telephone) brought onto a jobsite for the contractor's temporary use during construction.

Tenant. A person who rents property or land from a landlord; a Lessee.

Tenant at Sufferance. A holdover tenant that remains in possession with the permission of the lessor; also called Tenant at Will.

Tenant at Will. See Tenant at Sufferance.

Tenants in Common. Two or more persons who own land together without the right of survivorship.

Tender. 1. The person who tends the needs of a mason, such as a hod carrier, pack carrier, or wheelbarrow person. 2. An offer or bid.

Tendon. A concrete prestressing bar, cable, strand or wire.

Tenement. 1. Apartment dwelling. 2. An apartment building of minimum standards occupied by poorer families, usually in a city.

Tenements. The rights and privileges that accompany real property when it is conveyed.

Tennis Court. A level, quadrangular space divided by a low net, upon which tennis is played.

Tennis Net. A low net that divides a tennis court.

Tenon. A tongue-like protrusion on the end of a piece which is tightly fitted into a rectangular slot (mortise) in the side of another other piece.

Tenon Saw. A handsaw with fine teeth, a thin blade, and a stiffened back, for cutting tenons and other fine work.

Tensile Strength. Resistance to elongation; the greatest longitudinal stress a substance can bear without rupture or remaining permanently elongated; the pulling force necessary to break a given specimen divided by the cross sectional area; units are pounds per square inch;

Tensile Stress. A stress caused by stretching of a material.

Tension. Stress which tends to elongate a member.

Tension Ring. A structural element, forming a closed curve in plan, which is in tension because of the action of the rest of the structure; a concrete or masonry dome commonly has a tension ring.

Tension Wood. Abnormal wood found in leaning trees of some hardwood species and characterized by the presence of gelatinous fibers and excessive longitudinal shrinkage; tension wood fibers hold together tenaciously, so that sawed surfaces usually have projecting fibers, and planed surfaces often are torn or have raised grain; tension wood may cause warping.

Tepidarium. The room containing the tepid bath in Roman baths (thermae).

Term. The time period stated in a lease or mortgage.

Terminal. 1. A device for fastening and terminating an electrical conductor. 2. A final ending. 3. A terminus for trains or long-distance buses.

Terminal Adapters. Electrical fittings attached to the end of a conductor or to a piece of equipment for taking power from an outlet in a way for which it was not designed.

Terminal Block. A decorative element forming the end of a block structure.

Terminal Box. A metal electrical box, usually with a removable cover, that contains leads from electrical equipment ready for connection to a power source.

Terminal Reheat Handling System. A reheat system that is a modification of a single zone system; this system contains a cooling coil only (unless a preheat coil is needed); air is ducted to all spaces at the same temperature as in a single zone system, but at a temperature low enough to satisfy the temperature and humidity requirements of the zone with the highest cooling load; a terminal reheat box at each zone reheats the air as needed; a constant supply of air is supplied to the space served.

Termination for Cause. The right of an owner or contractor to end a construction contract if the other commits certain specified acts.

Termination for Owner's Convenience. The owner's right to end a construction contract for any reason to suit the owner's convenience.

Termite. A wood destroying social insect; Subterranean Termites live in the earth, and will build mud tunnels from the earth up to the wood structure; Kalotermes, also called Drywood Termites, swarm and will enter the structure where they will bore into the wood and hollow out the structural members.

Termite Control. The act or process of spraying chemicals into buildings and structures to control the spread of termites.

Termite Shield. A sheet metal barrier to discourage entry into the structure by subterranean termites.

Term Loan. A loan generally obtained from a bank or insurance company with a maturity greater than one year and less than 7 to 10 years.

Terms. Conditions of payment in a sale or purchase of merchandise.

Terne. An alloy of lead and tin, used to coat sheets of carbon steel, stainless steel, or copper for use as metal roofing sheet.

Terra Cotta. Hard baked clayware or tile, of variable color, averaging reddish red-yellow in hue and of high saturation.

Terra Sigillata. A porous red clay ware characterized by embossed decorations of the same color and a satin-like unglazed surface; originated on the Island of Samoa.

Terrazzo. A finish floor material consisting of concrete with an aggregate of marble chips selected for size and color, which is ground and polished smooth after curing; may also be precast and pre-finished.

Terrazzo Concrete. SeeTerrazzo.

Terrazzo Divider. A metal strip that separates areas of terrazzo.

Terrazzo Epoxy. A two-part adhesive, employing epoxy resin, an epoxy hardener used for bonding marble or other stone chips set in Portland cement to a backup material.

Terrazzo Floor. A high polished floor made from small marble or stone fragments embedded in a portland cement matrix.

Terrazzo Receptor. A shallow sink of a shower, forming its floor, made of Terrazzo.

Terrazzo Removal. The act or process of demolishing and discarding old terrazzo floors or walls or terrazzo tile.

Terrazzo Tile. A terrazzo surface, on a portland cement and sand body, made by a mixture of marble chips and portland cement and usually ground smooth.

Terrazzo Wainscot. A wainscot made of terrazzo or terrazzo tile wall facing.

Terrestrial. Of or relating to the earth.

Tertiary Color. A color produced by mixing two secondary colors or by mixing a primary color with a secondary color.

Tessera (pl. tesserae). A small chip of glass or marble used in mosaic formations.

Test. A trial, examination, observation, or evaluation used as a means of measuring the physical or chemical characteristics of a material, or a physical characteristic of a structural element or a structure.

Testate. One who dies leaving a will.

Test Cut. A sample of the roof membrane, usually 4 inches x 40 inches in size, that is cut from a roof membrane to determine the weight of the membrane, the number of plies, and the average interply bitumen poundages.

Testimony. Oral evidence offered by a witness in the course of a hearing, a trial, or a deposition.

Testing Laboratory. A place equipped to test construction materials to determine if they meet certain specified chemical, physical, and engineering criteria.

Testing Machine. A device for applying test conditions and accurately measuring results.

Testing Weld. The act or process of testing the strength of a weld.

Test Light. Light provided with test leads; used to test or probe electrical circuits to determine if they have electricity.

Test Pattern. Spray pattern used in adjusting spray gun.

Test Pile. A foundation pile which is driven before the final design is done to see what bearing is developed and what length of pile is actually needed; a pile which is tested by placing a predetermined load on it, commonly done by erecting a crib on the pile and then filling it with ballast.

Test Plug. A temporary plumbing fitting which is inserted into the end of a plumbing system to facilitate in pressure and leakage testing.

Test Set. Equipment used to check standards of water condition.

TEV. Thermostatic Expansion Valve.

Texture. 1.Visual or tactile surface characteristics. 2. A surface decoration. 3. In carpet, a surface effect obtained by using different heights of pile or two or more forms of yarn, or by alternating the round and cut pile wires, by brocade engraving, simulated or actual carving or shaving with an electric razor, or other special treatment of the design, to give added interest beyond that provided by the woven design or tones.

Texture Paint. One which may be manipulated by brush, trowel or other tool to give various patterns.

TFS. Texas's Forest Service.

Theater Equipment. Props, lighting, curtains, stage platforms and other devices and objects used in a theater.

Theater Stage. A raised platform where plays, musicals or other theatrical events are performed.

Theodolite. A surveyor's instrument for measuring vertical and horizontal angles; a transit.

Theoretical. In spot, seam and projection welding, the force, neglecting friction and inertia, available at the electrodes of a resistance-welding machine by virtue of the initial force application and the theoretical mechanical advantage of the system.

Therm. A heat unit; 100,000 BTUs.

Thermae. Roman public baths; see Calidarium.

Thermal. Pertaining to heat.

Thermal Capacity. 1. A measure of how much energy is needed to heat up a substance. 2. The quantity of heat needed to warm a collector up to its operating temperature.

Thermal Conductance (C). A unit of heat flow that is used for specific thicknesses of material or for materials of combination construction, such as laminated insulation. The formula for thermal conductance is. C = k times the thickness in inches

Thermal Conductivity (k). The heat energy that will be transmitted by conduction through one square foot of one inch thick homogeneous material in one hour when there is a difference of 1 degree Fahrenheit perpendicularly across the two surfaces of the material. The formula for thermal conductivity is. k = BTU/Square Foot/Inch/Hour/Degree Farenheit.

Thermal Cutout. An overcurrent protective device which contains a heater element in addition to and affecting a renewable fusible member which opens the circuit; it is not designed to interrupt short circuit currents.

Thermal Fracture. A compression crack caused by expansion of peripheral building components.

Thermal Inertia. The tendency of a building to remain at the same temperature or to fluctuate only very slowly.

Thermal Insulation. Material used to retard the flow of heat through an enclosing surface.

Thermal Insulator. An insulation material that lowers the thermal conductivity of a part of a building system, such as a floor, wall, ceiling, or roof assembly.

Thermal Lag Factor. The time interval between the time heat enters a material at one side and exits at the other side.

Thermal Movement. The expansion and contraction of any material caused by temperature differences.

Thermal Protector. An electrical device that automatically opens a circuit in the event of overheating, thereby protecting the appliance from possible damage.

Thermal Radiation. Electromagnetic radiation emitted by a warm body.

Thermal Relay. Heat operated electrical control used to open or close a refrigeration system electrical circuit; this system uses a resistance wire to convert electrical energy into heat energy; also called Hot Wire Relay.

Thermal Resistance (R). An index of a material's resistance to heat flow; it is the reciprocal of thermal conductivity (k) or thermal conductance (C). The formula for thermal resistance is.

Thermal Shock. 1. Stress built up by sudden and appreciable changes in temperature. 2. The shock produced and also the stress produced resulting from sudden temperature changes in a roof membrane; for example, when a rain shower follows brilliant sunshine. 3. A stress created by an extreme change in temperature that may result in cracking of the plaster which has not yet attained its ultimate strength.

Thermal System Insulation. Insulative material applied to pipes, fittings, boilers, tanks, ducts or other interior structural components to prevent heat loss or gain or water condensation.

Thermal Transmission Value. The resistance factor to the conductance of heat.

Thermistor. A semiconductor which has electrical resistance that varies with temperature.

Thermocouple. A pair of dissimilar metals create a thermoelectromotive force that can be measured and converted to temperature differences.

Thermocouple Thermometer. Electrical instrument using thermocouple as source of electrical flow, connected to a milliammeter calibrated in temperature degrees.

Thermodisk Defrost Control. Electrical switch with bimetal disk controlled by temperature changes.

Thermodynamics. The science dealing with the relationship between heat and other forms of energy.

Thermoelectric Refrigeration. Refrigerator mechanism that depends on Peltier effect; direct current flowing through electrical junction between unlike metals provides heating or cooling effect depending on direction of flow of current.

Thermometer. Device for measuring temperatures.

Thermomodule. A number of thermocouples used in parallel to achieve low temperatures.

Thermopane Window. Tradename; atype of glass constructed in a manner to protect against sound, heat, heat loss or moisture; double or triple glazed glass.

Thermopile. Number of thermocouples used in series to create a higher voltage.

Thermoplastic. Having the property of softening when heated and rehardening when cooled.

Thermoplastic Glues and Resins. Glues and resins that are capable of being repeatedly softened by heat and hardened by cooling.

Thermosetting. Material which undergoes irreversible chemical reaction when heated and molded; once formed, it cannot be reheated and reshaped.

Thermosetting Glues and Resins. Glues and resins that are cured with heat and do not soften when subjected later to high temperatures.

Thermostat. Device which senses ambient temperature conditions and, in turn, acts to control a circuit.

Thermostat Cable. A specific electrical system cable which operates an automatic device for regulating the temperature in a room, space, or area.

Thermostatic Control. Device which operates system or part of system based on temperature change.

Thermostatic Expansion Valve (TEV). Control valve operated by temperature and pressure within evaporator. It controls flow of refrigerant; control bulb is attached to outlet of evaporator.

Thermostatic Motor Control. Device used to control cycling of unit through use of control bulb; bulb reacts to temperature changes.

Thermostatic Valve. Valve controlled by temperature change response elements.

Thermostatic Water Valve. Valve used to control flow of water through system, actuated by temperature difference; used in units such as water-cooled compressor and/or condenser.

Thickness. 1. The smallest of the three dimensions of a solid object. 2. The quality of being thick. 3. A layer of material. 4. The dimension at right angles to the face of the wall, floor, or other assembly in which masonry units are used.

Thin Coat. A one coat plaster system over gypsum board.

T Hinge. A surface hinge with the short member attached to the jamb and the long member attached to the door.

Thin Set. The bonding of tile with suitable materials, applied approximately 1/8" thick.

Thin Shell. A three dimensional spatial structure made up of one or more curved concrete slabs or folded plates whose thicknesses are small compared to their other dimensions; thin shells are characterized by their three dimensional load carrying behavior which is determined by the geometry of their forms, by the manner in which they are supported, and by the nature of the applied load.

Thin Wall Conduit. See EMT Conduit.

Thincoat-High Strength Plaster. See Veneer Plaster.

Thingamabob. See Thingamajig.

Thingamajig. Something that is difficult to classify or whose proper name is unknown or forgotten; a Thingamabob.

Thinner. A volatile liquid added to finishing material to make it flow more easily and smoothly; also called a Diluent.

Thinners. Volatile liquids used to reduce the viscosity of paint and varnish; also called Diluents.

Third Party Beneficiary. A person who is not a party to a contract but who is an intended beneficiary thereof and may therefore enforce it.

Thixotropic Paint. Property exhibited by some paint of becoming fluid when shaken or disturbed; after cessation of mechanical disturbance, such as stirring or putting brush into paint, rigidity develops again.

Thixotropy. The property of various gels of becoming fluid when disturbed, as by shaking or stirring.

Thomas Jefferson Ideas. The advocacy of grid pattern planning where frequent public squares are left open between building blocks.

Threaded Coupling. A fitting for joining two lengths of threaded pipe.

Threading. The process of forming external screw threads on a rod or pipe.

Threadless Coupling. A fitting for joining two lengths of pipe fastened by soldering or cementing.

Threads. Spiral ridges on a member, like a pipe, by which parts can be screwed together.

Three Coat Plastering. The application of plaster in three successive coats, leaving time between coats for setting and drying of the plaster.

Three Dimensional. Having, or appearing to have, three dimensions, height, width, and depth.

Three-Phase. Operating by means of combination of three alternating current circuits which differ in phase by one-third of a cycle.

Three Point Lock. A device to lock a door to the jamb at three points.

Three Quarter. A brick with one end broken off; a three quarter brick.

Three-Quarter Bath. A room containing a toilet, lavatory, and shower.

Three-Way Switch. A switching circuit enabling a light fixture to be controlled from two points.

Three-Way Valve. Multi-orifice flow control valve with three fluid flow openings.

Threshold. 1. The wood or metal beveled floor piece at door openings which commonly separates non-continuous floor types. 2. A strip of wood or metal beveled on each edge and used above the finished floor under outside doors; also called a Sill or a Saddle.

Threshold, Door. A beveled piece of floor trim over which a door swings.

Threshold Limit Value (TLV). Maximum concentration of solvent vapor in parts per million parts of air in which a worker may work eight consecutive hours without an air fed mask; the lower the TLV number the more toxic the solvent.

Throat. The opening at the top of a fireplace, through which the smoke passes into the smoke chamber and chimney; the damper is placed in the throat.

Throne. A chair of state for a sovereign or bishop.

Throttling. Expansion of gas through an orifice or controlled opening without gas performing any work as it expands.

Through Dry. Ability of film to show no loosening, detachment or evidence of distortion when the thumb, placed on film with maximum arm pressure, is turned through 90 degrees in the plane of the film.

Through-Wall Flashing. A sheet metal sleeve through a parapet wall, integrated with the abutting counterflashing and roofing to prevent water from intruding.

Throw. The distance that a door latch or bolt may be fully extended.

Thrust. A lateral or inclined force resulting from the structural action of an arch, vault, dome, suspension structure, or rigid frame.

Thrust Block. A concrete mass around water piping to prevent movement caused by the sudden change in velocity of water flow.

Thrust Fault. A geological fault caused by compression in which the upper older layers of rock are thrust over the lower younger ones at a very low angle.

Thumb Screw. A screw with a knurled or flat-sided head that may be grasped between the thumb and forefinger for turning.

Thumbnail Proof. Checking hardness of a finish by pressing thumbnail against it.

Thumb Tack. A tack with a broad flat head so it may be pressed into a surface with the thumb.

Thyristor. An electrical semiconductor device for converting alternating current into direct current.

Ticky-Tacky. 1. Sleazy or shoddy material. 2. Cheap, repetitive housing built of shoddy materials.

Tie. 1. A device for holding components together, a structural device that acts in tension. 2. A stirrup tie.

Tie Bars. 1. Steel reinforcing bars at right angles and tied to main reinforcement to keep it in place. 2. Bars extending across a construction joint.

Tie Beam. A reinforced concrete beam cast as part of a masonry wall, whose primary purpose is to hold the wall together, especially against seismic loads, or cast between a number of isolated foundation elements to maintain their relative positions.

Tie Beam. See Collar Beam.

Tied Column. A square or rectangular reinforced concrete column containing longitudinal reinforcing and lateral ties.

Tie Elements. Elements which serve to transmit inertia forces and prevent separation of such building components as footings and walls.

Tie-In. The point or line of interface of new materials with old in alterations or repairs.

Tie, Masonry. See Masonry Tie.

Tie, Mesh. See Mesh Tie.

Tier. 1. A row or rank or unit of a structure, as one of several placed one above another. 2. One of the vertical layers in the thickness of a masonry wall.

Tie Rod. A steel rod that acts in tension and commonly holds together wall forms while concrete is being poured.

Ties. The attaching of metal lath to furring channels by use of light gauge malleable wire; see Butterfly Tie, Saddle Tie, and Stub Tie.

Tie Strap. A metal plate that fastens two parts together as a post, rod, or beam.

Tie Wire. 1. Light gauge wire used as temporary fasteners in construction work. 2. Soft annealed steel wire used to join lath supports, attach lath to supports, and attach accessories. 3. Wire (generally No. 16, No. 15 or No. 14 gauge) used to secure intersections of reinforcing bars for the purpose of holding them in place until concreting is completed.

TIG Welding. See Gas Tungsten-Arc Welding.

Tile. 1. To install ceramic tiles, flooring tiles, roofing tiles, or drainage tiles. 2. A ceramic surfacing unit, usually relatively thin in relation to facial area, made from clay or a mixture of clay and other ceramic materials, called the body of the tile, having either a glazed or unglazed face and fired above red heat in the course of manufacture to a temperature sufficiently high to produce specific physical properties and characteristics. 3. A fired clay product that is thin in cross section as compared to a brick; either a thin, flat element (ceramic tile or quarry tile), a thin, curved element (roofing tile), a hollow element with thin walls (masonry tile), or a pipe-like shape (drainage tile).

Tile, Acoustical. See Acoustical Tile.

Tile Adhesive. Organic adhesive used for bonding tile to a surface; rubber solvents and resin-based and rubber emulsions can be used as adhesives.

Tile Assemblies. Tile assembled into units or sheets by suitable material to facilitate handling and installation; tile may be face-mounted, back-mounted or edge-mounted.

Tile, Back Mounted. See Back Mounted Tile.

Tile, Clay. See Clay Tile.

Tile, Conductive. See Conductive Tile.

Tile Cutter. Any device that cuts tile; one of the most efficient and economical tools is the adjustable hand-drawn tile cutting board.

Tile, Face Mounted. See Face Mounted Tile.

Tile Mounted. Tile fastened together in sheets to facilitate handling and setting.

Tile, Quarry. See Quarry Tile.

Tile, Resilient. See Resilient Flooring.

Tile Setting. The skilled trade of installing ceramic tile on floor, wall, stairs, or counter surfaces.

Tile, Sewer. See Sewer Tile.

Tile, Terrazzo. See Terrazzo Tile.

Tilt Angle. The angle that a flat solar collector surface forms with a horizontal.

Tilt-Up Construction. A method of constructing concrete walls in which panels are cast and cured flat on the floor slab, then tilted up into their final positions.

Tilt-Up Wall. See Tilt-Up Construction.

Timber. Lumber with cross-section over 4 by 6 inches, such as posts, sills, and girders.

Timber Framing. Structural framing by the use of timber for load carrying applications.

Timber Parking Barrier. A barrier made from thick wood to prevent the encroachment of vehicles.

Timber, Standing. Timber still on the stump.

Time and Material. A method of paying a contractor for work, consisting of reimbursement for time expended and cost of materials.

Time Clock. A clock that stamps an employee's starting and ending times on a time card.

Timekeeper. A clerk who keeps records of the time worked by employees.

Timer. Clock-operated mechanism used to control opening and closing of an electrical circuit.

Timer-Thermostat. Thermostat control which includes a clock mechanism; unit automatically controls room temperature and changes temperature range depending on time of day.

Time-Sharing Ownership. A method of sharing in the ownership, usually of recreational properties, in which each co-owner has the right to occupancy at specified times only.

Time-Weighted Average. In air sampling, this refers to the average air concentration of contaminants during a particular sampling period.

Tin. A soft faintly bluish white lustrous low-melting crystalline metallic element that is malleable and ductile at ordinary temperatures and that is used as a protective coating, in tinfoil, and in soft solders and alloys.

Tinge. Slight trace of color.

Tinner. A tinsmith.

Tinning. In soldering, coating the metals to be joined with a thin layer of solder.

Tin Shingle. A small piece of tin used in flashing and repairing a shingle roof.

Tinsmith. A worker in tin.

Tin Snip. A hand tool for shearing tin.

Tint. 1. A hue lightened by addition of white; a color produced by adding white pigment or paint to a colored pigment or paint, with the amount of white greater than the amount of colored pigment.

Tinted Glass. Glass that has been treated to reduce transmitted glare.

Tissue Dispenser. A container with a slim opening that holds tissues and allows one to be drawn at a time.

Titania Porcelain. A vitreous ceramic whiteware for technical application in which titania, TiO_2, is the essential crystalline phase.

Titanium. A metal which is the basis for the pigment, titanium dioxide.

Titanium Calcium. Paint pigment made by combining titanium dioxide of rutile type and calcium sulphate.

Titanium Dioxide. White pigment used extensively in paint making; comes in two forms, rutile and anatase; it is chemically inactive and is not affected by dilute acids, heat or light.

Title. The right to ownership of property.

Title Deed. Evidence if ownership of property.

Title Insurance. A guarantee of title issued by an insurance company.

Title Search. A research of the county records to determine the condition of title of a parcel of real property.

TL. Transmission Loss.

TLV. Threshold Limit Value.

TMS. The Masonry Society.

Toeboard. Wood or metal vertical metal member at floor or landing level placed at floor or landing edge to prevent objects from falling or persons from stepping off the edge.

Toe Nail. Nailing at an angle, as from the bottom of a stud to the sole plate.

Toe Plate. 1. A metal bar or plate fastened to the outer edge of a floor grating. 2. A metal plate fastened to the rear of a stair tread. 3. A protective metal plate fastened to the bottom rail of a door.

Toe Space. A recess at the bottom of a cabinet supporting a work counter, to provide space for the worker's toes.

Toggle Bolt. A bolt and nut assembly to fasten objects to hollow construction assembly from only one side. Nut has pivoted wings that close against spring when nut end of assembly is pushed through hole and spring open on other side in void of construction assembly.

Toilet. A bathroom or washroom fixture that serves as a receptacle for human waste; also called a Water Closet.

Toilet Bowl. The oval part of a toilet which receives the waste and fills with water after flushing the toilet tank.

Toilet Carrier. A metal structure attached to a bathroom wall to support and mount a wall-hung toilet.

Toilet Partition. Privacy panels in a toilet enclosure.

Toilet Room. A room containing one or more toilets and usually one or more lavatories.

Toilet Room Accessories. Toilet room equipment such as toilet paper dispensers, paper towel dispensers, hand dryers, sanitary napkin dispensers, grab bars, and grab rails.

Toilet Seat and Cover. A hinged seat for a toilet and its hinged cover.

Toilet Tank. The reservoir located in the back of a wate. Tool used to cut internal threads.

Tolerance. 1. The allowable deviation from a standard. 2. Extra space to allow for dimensional differences.

Toluene. Lacquer diluent made normally by coal tar distillation; also called Toluol.

Toluidine Red. Brilliant non-bleeding red pigment made from coal tar product.

Toluol. See Toluene.

Tomb. 1. A grave; place of entombment. 2. A house, chamber, or vault for the dead.

Ton. 1. A measure of weight equaling 2000 pounds. 2. The metric ton is 1000 kilograms or 2200 pounds.

Tone. 1. A sound of only a single frequency...2. A graduation of color, either a hue, a tint, or a shade, as a gray tone.

Tone Down. The process of reducing visual prominence of an installation by the application of external coatings; blending of overall color scheme with the surrounding environment.

Tone-on-Tone. Carpet pattern made by using two or more shades of the same hue.

Toner. A color modifier.

Tongue. 1. A projecting edge, as on a board, that fits into a groove of another piece, as in a tongue and groove joint. 2. See Spline. 3. See Steel Square.

Tongue and Groove (T&G). A type of lumber, metal, or precast concrete having matching or mated edges to provide a tight fit.

Ton of Refrigeration. Refrigerating effect equal to the melting of 1 ton of ice in 24 hours; this may be expressed as. 288,000 Btu/24 hours or 12,000 Btu/hour or 200 Btu/minute.

Too Much Drag. Refers to paint that has excessive pull or drag in its application.

Tool Belt. A webbed or leather belt worn by carpenters, electricians, and other artisans from which frequently used hand tools are slung.

Tooled Joint. A masonry joint where the surface has been tooled to compress the mortar; the exposed surface can be flat, concave, or v-shaped, depending on the configuration of the tool.

Tooling. Compressing and shaping the face of a mortar joint with a special tool other than a trowel; also called jointing; See Jointing, 3.

Tooth. 1. Profile; mechanical anchorage; surface roughness. 2. Roughened or absorbent quality of a surface which affects adhesion and application of a coating.

Toothed Ring. A shear-resisting timber connector used in the manufacturing of large member wood trusses.

Toother. A brick projecting from the end of a wall against which another wall will be built.

Toothing. The system of the construction at the end of a wall in which every other course projects one half of a brick length; then another wall may be tied into this staggered brick end.

Toothing In. The joining of a new masonry wall on to the old toothed wall.

Top Bars. Steel reinforcing bars near the top of reinforced concrete.

Top Chord. The top flat or sloping member of a truss

Top Coating. Finish coat.

Top Colors. Colors of the yarn used to form the design, as distinguished from ground color.

Top Dip. The highest water or waste point in the bottom section of a trap.

Top Mopping. The finish coat of hot bitumen on a built up roof.

Topographic Map. See Topographic Survey.

Topographic Survey. A land survey that shows topography and all other physical features; also called a Topographic Map or a Contour Map.

Topography. 1. The form of the terrain. 2. The description of the surface features in graphic terms, depicted by contour lines showing ground elevation.

Topology. The study of the geometric properties and spatial relations of flat and solid shapes that are unchanged by squeezing, stretching, twisting, or changing of size.

Top Plate. The horizontal member at the top of a stud wall, usually supporting rafters.

Topping. Fine material forming a surface layer or dressing for a road or grade.

Topping Compound. A specially formulated gypsumboard joint compound designed for the final joint finishing coat; not to be used to embed tape.

Topping, Concrete. See Concrete Topping.

Topping, Granolithic. See Granolithic Topping.

Top Set Base. Vinyl or rubber base with an integral cove, cemented to the wall, set on top of resilient floor coverings.

Topsoil. Surface soil at and including the average plow depth, soil which is used as a planting or growing medium.

Top Weir. The top part of the outlet of a trap.

Torch Brazing. A brazing process wherein coalescence produced by heating with a gas flame and by using a nonferrous filler metal having a melting point above 800° F. but below that of the base metal; the filler metal is distributed in the joint by capillary attraction.

Torching. Application of direct flame to a membrane for the purpose of heating, melting, or adhering.

Tornado. A violent storm of whirling winds of high speed, over a narrow path often accompanied by a funnel shaped cloud.

Torque. 1. Turning or twisting action. 2. Moment.

Torque, Full Load. See Full Load Torque.

Torque, Stall. See Stall Torque.

Torque, Starting. See Starting Torque.

Torque Wrench. Wrench which may be used to measure torque or pressure applied to a nut or bolt.

Torsion. The rotation of a diaphragm caused by lateral forces and whose center of mass does not coincide with the center of rigidity.

Tort. A negligent or intentional wrongful act that damages the person or property of another, the wrongful nature of which is independent of any contractual relationship.

Torus. A geometric solid shaped like a doughnut.

Total Absorption. The amount of water a masonry unit will absorb when immersed in water.

Total Alkalinity. Actual amount of alkali salts present in swimming pool water.

Total Cost. A method of computing damages sustained by a contractor because of breaches of contract causing the contractor to operate in an inefficient or unproductive manner.

Total Design Displacement. The design-basis of earthquake lateral displacement, including additional displacement due to actual and accidental torsion, required for design of the isolation system, or an element thereof.

Total Heat. Sum of both the sensible and latent heat.

Total Maximum Displacement. The maximum capable earthquake lateral displacement, including additional displacement due to actual and accidental torsion, required for verification of the stability of the isolation system, or elements thereof, design of building separations, and vertical load testing of isolator unit prototype.

Total Static Head. Static head from the surface of the supply source to the free discharge surface.

To the Weather. A term applied to any part of the structure which faces the elements; a shingled roof is to the weather, but the framing system is not.

Touch Up Primer. Field touch-up of prime painting for steel members after erection.

Touch Up Painting. Spot repair painting usually conducted after initial painting.

Toughness. A quality of a material which permits it to absorb a relatively large amount of energy, to withstand repeated shocks, and to undergo considerable deformation before yielding or breaking.

Towel Bar. A horizontal rod, usually secured to a wall, on which a towel can be hung or laid.

Towel Dispenser. A container that holds paper towels for future use, and which has an opening allowing one towel to be removed at a time.

Tower. A tall structure, constructed of frames, braces, and accessories rising to a greater height than the surrounding area.

Tower, Cooling. See Cooling Tower.

Townhouse. A single-family house of two or three stories, often connected to similar houses in a row.

Township. In land ownership and legal description, a township is a square 6 miles on a side, totaling 36 square miles; each square mile is called a section.

Toxic. Poisonous.

Toxicity. Degree of poisonousness or harmfulness.

Toxins. A poison produced by a living organism.

TPI. Truss Plate Institute.

Trace. 1. A very small, barely perceptible, amount or vestige of a material. 2. To delineate the outline of an object. 3. To copy a drawing by use of an overlay of thin paper.

Tracer Gases. Compounds, such as sulfur hexafluoride, which are used to identify suspected pollutant pathways and to quantify ventilation rates; tracer gases may be detected qualitatively by their odor or quantitatively by air monitoring equipment.

Tracery. 1. Ornamental stone openwork in the upper part of a gothic window. 2. A fine decorative pattern.

Tracheid. The elongated cells that constitute the greater part of the structure of the softwoods (frequently referred to as fibers); also present in some hardwoods.

Tracing. A reproducible drawing made on semitransparent paper called tracing paper; such a drawing on any kind of semitransparent drawing medium.

Tracing Paper. Transparent drafting medium from which photosensitive prints can be made of the drawing.

Track. 1. A metal way serving as a guide for a movable lighting fixture. 2. Metal channel; see Runner. 3. A specially laid out course, as for racing.

Track Lighting. An assembly with a U-shaped electrified member attached to a ceiling or wall, acting as a channel for sliding light fixtures.

Tract. A defined area of land.

Traction. The friction grip of a wheel or track on the surface on which it moves.

Trade Discount. A discount or deduction from a catalog or list price extended to the trade or quantity buyer.

Trade Union. An organized association of workers in a trade, group of trades, formed to protect and further their rights and interests.

Traffic. 1. The movement of people or vehicles through an area. 2. The passing to and fro of persons with special reference to carpet wear resulting therefrom.

Traffic Deck. A walking surface placed on top of a roof membrane.

Traffic Paint. A paint, usually white, red or yellow, used to designate traffic lanes, safety zones, and intersections.

Traffic Sign. A posted command, warning, or direction for motor vehicle drivers or pedestrians.

Trailer. A vehicle designed to be hauled or to serve parked as a dwelling or place of business.

Trammel. A Beam Compass.

Trammel Bar. A tile layout tool used to erect perpendicular lines and to bisect angles.

Transducer. Device turned on by change of power from one source for purpose of supplying power in another form to second system.

Transept. Either arm of the part of a cross-shaped church at right angles to the nave.

Transfer. That act of transferring stress in prestressing tendons from jacks or pretensioning bed to concrete member.

Transfer Switch. A switch in an electrical system that transfers the load to another circuit when the voltage drops below a predetermined level.

Transformer. A device that changes, or transforms, alternating current from one voltage to another.

Transformer Pad. A precast concrete block or stone placed under a transformer to spread and support its weight.

Transformer-Rectifier. Combination transformer and rectifier in which input a-c current may be varied and then rectified into d-c current.

Transistor. An electronic component that switches or amplifies a current; a semiconductor device with three connections.

Transit. See Theodolite.

Transit Mixed Concrete. Concrete produced from a central batching plant, where the materials are proportioned and placed in truck mixers for mixing enroute to the job or after arrival there.

Transition. 1. A passing or change from one place, state, or condition to another. 2. The physical part needed to effect the change, such as a piping reducer.

Transition Lot. A lot in a subdivision that consists of both cut and fill.

Transition Point. In a swimming pool, the point at which the floor slope changes from shallow to deep area.

Transition Primer. Coating compatible with primer and with a finish coat, though the latter is not compatible with the primer; also called Barrier Coating or Block Coating.

Translate. 1. To move from one place to another, without rotation. 2. Motion of a body along a line without rotation or turning.

Translation. Motion of a body along a line without rotation, or turning.

Translucent. Permitting the passage of light.

Translucent Panel. A building panel that permits the passage of light but not vision.

Transmission Factor. See Transmittance.

Transmission Loss (TL). The reduction of airborne sound power that is caused by placing a barrier, wall or material, between the reverberant sound field of a source and its receiver; transmitted loss is a property of the barrier.

Transmittance. The ratio of light transmitted through a material to the total incident light falling on it; also called Transmission Factor or T..

Transom. 1. A transverse piece in a structure, lintel. 2. A horizontal crossbar in a window, over a door, or between a door and a window or a fanlight above it. 3. A window above a door or other window built on and commonly hinged to a transom.

Transom Bar. The part of a door or window frame that separates the top of a door or window from the bottom of a transom light or sash.

Transom Catch. A fastener applied to a transom sash and having a ring by which the latch bolt is retracted.

Transom Chain. A short length of chain used to limit the opening of a transom sash, usually with a plate provided at each end for attachment.

Transom Lift. A vertically operated device attached to a door frame and transom sash by which the transom sash may be opened or closed.

Transom Sash. The framework of a glass window over a door or window; may be hinged for ventilation.

Transparent Mirror. A mirror with a transparent coating that allows vision from the space on the darker side to the space on the more brightly lit side; from the brighter side, it reflects like a mirror; also called One-Way Glass.

Transudate. A material that has transuded or seeped.

Transude. To pass through a membrane or permeable substance; to seep or exude.

Transverse. At right angles to the long direction of the member; crosswise; also referred to as lateral.

Transverse Strength. A standard measurement of the relative flexural strength of gypsum board products conducted in accordance to ASTM C 473; expressed as pounds force or Newtons.

Trap. Located at a plumbing fixture waste line, designed to hold a quantity of water that provides a seal to prevent gasses in the sewer system from entering a room.

Trapdoor. A door or hatch in a floor, ceiling, or roof.

Trap, Float. See Float Trap.

Trap, Grease. See Grease Trap.

Trap Seal. The vertical distance between the crown weir and the top dip of the trap.

Trap Vents. In all traps it is necessary that a back vent connection be made to the drain or waste pipe on the sewer side of the trap; if the back vent is not provided, the trap could be emptied by siphonage.

Trash Chute. A device either constructed on the inside of a building or structure or hung outside for the removal of waste materials from upper floors.

Trash Compactor. A device that compresses waste materials to conserve storage space.

Trash Handling. The act or process of transporting or removing waste materials.

Traveling Crane. A tower crane mounted on tires, rails, or crawlers.

Traverse Rod. A device that draperies or curtains are hooked on so they can be opened or closed.

Travertine Floor. A floor constructed of richly patterned, marble-like limestone.

Travertine. A richly patterned, marble-like form of limestone.

Tray, Cable. See Cable Tray.

Tread. The horizontal surface of a step in a stair.

Tread, Grating. See Grating Tread.

Treated Lumber. Lumber infused or coated with stain or chemicals to retard fire, decay, insect damage, or deterioration due to weather.

Treated Pole. Wooden pole infused or coated with stain or chemicals to retard fire, decay, insect damage, or deterioration due to weather.

Treated Post. Vertical post infused or coated with stain or chemicals to retard fire, decay, insect damage, or deterioration due to weather.

Tree Ball. See Root Ball.

Tree Cutting. The trimming of leaves, branches or larger portions of a tree, or the chopping down of a tree.

Tree Maintenance. The pruning or trimming of certain parts of a tree for aesthetic or practical purposes.

Tree Moving. The digging out of a tree with its root system intact, to relocate it to a different place.

Treenail. A wooden pin, peg, or spike used chiefly for fastening planking and ceiling to a framework.

Tree Removal. The chopping down, carrying off, or disposal of a tree or tree segments.

Tree Trimming. The pruning or light cutting of branches, leaves or other portions of a tree for practical or aesthetic purposes.

Tree Well. A planting space for one or more trees.

Trefoil. A three cusped element in Gothic tracery.

Trellis. A light latticework on which trees or shrubs may be trained.

Tremie. A large funnel with a tube attached, used to deliver concrete into deep forms or beneath water or slurry; a tremie slows down the concrete and resists segregation of the aggregates.

Trench. A long, narrow excavation; a ditch.

Trencher. A mechanical device used to dig narrow channels in the ground.

Trench Frame and Cover, Cast Iron. Frame and cover of cast iron components to cover recessed lineal element designed to collect water from surface and deliver to drainage pipe or pit.

Trenching. The act or process of digging narrow channels in the ground.

Trespass. Unauthorized entry upon the real property of another.

Trestle. 1. A braced framework bridge supporting a railroad track. 2. A table frame consisting of a horizontal member supported by a pair of divergent legs at each end.

Triad Color Harmony. Harmony obtained by using colors from three equidistant points of the color wheel; red, yellow, and blue make up a triad.

Trial. A hearing conducted by a judge at which litigants are entitled to introduce evidence and arguments supporting, and defending against, claims.

Triangle. 1. Three sided plane figure; a polygon of three sides. 2. A transparent plastic drafting tool in the shape of a triangle, usually either 45 degrees, 30-60 degrees, or adjustable.

Trial Batch. A batch of concrete prepared to establish or check proportions of the constituents.

Tribology. The science dealing with friction, wear, and lubrication of interacting surfaces.

Trichloromonofluoromethane. Low pressure, synthetic chemical refrigerant which is also used as a cleaning fluid.

Trichloromonofluoroethane. Complete name of R-113, a synthetic, nontoxoic and nonflammable Group 1 refrigerant in rather common use; chemical components which make up this refrigerant are chlorine, fluorine, and ethane.

Trig. The bricks laid in the middle of a wall between the two main leads to overcome the sag in the line and also to keep the center plumb in case there is a wind bearing upon the line.

Triglyphs and Metopes. Parts of a Doric frieze; the triglyphs are vertical lines and the metopes are the spaces between.

Trigonometry. The branch of mathematics dealing with the relations of the sides and angles of right triangles and with the relevant functions of any angles.

Trilinear. Involving three lines

Trim. Finish materials, such as moldings applied around openings (window trim, door trim) or where walls join the floor or ceiling of a room (baseboard, cornice, and other moldings).

Trim Enamel Paint. Surface coating differing from ordinary house paint by faster drying, by having more gloss and showing fewer brush marks; used mostly on trim, shutters, and screens.

Trimmer. 1. A beam or floor joist into which joists are framed. 2. A beam or joist to which a header is nailed in framing, as for a chimney or a stairway.

Trimmer Arch. An arch, usually of brickwork and of low rise used for supporting the fireplace hearth.

Trimmers. Tile units of various shapes consisting of such items as bases, caps, corners, moldings, and angles necessary or desirable to make a complete installation and to achieve sanitary purposes as well as architectural design for all types of tile work.

Trimming. Putting the inside and outside finish and hardware on a building.

Trim Painting. The act or process of applying paint to trim work.

Triple Glazing. An assembly of three panes of glass separated by air spaces; triple glass.

Triple Oven. A mechanical cooking device that has three separate chambers with doors.

Triple Point. The condition of temperature and pressure under which the gaseous, liquid, and solid phases of a substance can exist in equilibrium.

Triplex. 1. A three-family house. 2. A Triplex Apartment.

Triplex Apartment. An apartment having rooms on three floors.

Tripod. A three legged support for something, such as a transit or camera.

Tripoli. See Rottenstone.

Triptych. A picture or carving in three panels.

Trisect. 1. To divide into three parts. 2. To divide into three equal parts.

Trisodium Phosphate (TSP). A crystalline compound that forms the base for cleaning compounds; especially good for washing walls prior to repainting.

Troffer. A channel-like enclosure for light sources.

Trombe Wall. A masonry wall, usually glazed on the outside, used to trap solar heat, as a part of a passive solar heating system.

Troposphere. Part of the atmosphere immediately above the earth's surface in which occur most weather disturbances.

Troughing. Making repeated dozer pushes in one track, so that ridges of spilled material hold dirt in front of the blade.

Trowel. A hand tool used by masons and plasterers to apply, shape, spread, or smooth plastic materials such as plaster and mortar.

Trowel Finish. The smooth finish coat surface of concrete, plaster, or stucco produced by troweling.

Trowelling. Applying, spreading, or smoothing plastic materials such as mortar, concrete, or plaster, using a hand trowel.

Truck. 1. A wheeled vehicle for moving heavy articles. 2. A strong automotive vehicle for hauling. 3. A small wheelbarrow consisting of a rectangular frame having at one end a pair of handles and at the other end a pair of small, heavy wheels and a projecting edge to slide under a load.

Truck Crane. A mechanical device mounted on the bed of a truck for hoisting or lifting materials.

Truck Mixer. A concrete mixer capable of mixing concrete in transit when mounted on a truck chassis.

Truck, Refrigerated. See Refrigerated Truck.

Truck Shelter. An enclosure or structure for the storage of motorized trucks.

Truck Well. A ramped-down depressed area adjacent to a loading door to allow trucks to be loaded or unloaded with the truck bed level with the building floor.

True Complement Color Harmony. Two colors directly across the color wheel from each other are true complements; examples are red and green, and orange and blue.

Truncate. To shorten by cutting off the top of an object.

Truss. 1. A triangular arrangement of structural members that reduces nonaxial forces on the truss to a set of axial forces in the members. 2. Structural framework of triangular units for supporting loads over long spans.

Trussed Bars. Steel reinforcing bars bent up to act as both top and bottom reinforcement.

Truss Panel. A chord segment between two successive joints of a truss.

Truss Reinforcing. The addition of cross bracing or other bracing members to stiffen and strengthen a framework of trusses.

Truss Steel. Steel framework of triangular units for supporting loads over long spans.

Truss, Wood. See Wood Truss.

Trust Deed. A document evidencing a security interest in real property; used in some states in place of a mortgage.

Trustee. In a trust deed, the entity to whom title to the property is legally entrusted for the benefit of the beneficiary (the lender).

Trustor. In a trust deed, the entity that has pledged property to guarantee repayment of a loan.

TSG. Tapered Steel Girder.

TSI. Thermal System Insulation.

TSP. Trisodium Phosphate.

T-Square. A drafting tool that, used with a drawing board, can be used to produce parallel lines on a drawing.

T-Stat. Thermostat.

Tsunami. A sea wave produced by a large displacement of the ocean bottom, the result of earthquakes or volcanic activity.

Tub, Bath. A basin with a drainage unit and water supply used primarily for bathing or washing purposes.

Tube. 1. A duct or pipe to convey liquids. 2. A structural load-bearing member in the form of a square or round pipe. 3. A tunnel. 4. Electron tube. 5. Vacuum tube. 6. Television Tube.

Tube, Constricted. Tubing reduced in diameter.

Tub Enclosure. An enclosure of glass or plastic mounted on a bathtub rim with sliding, folding, or swinging door panels, to keep shower spray in the compartment.

Tube-In-Plate Absorber. An aluminum or copper sheet metal absorber plate in which the heat transfer fluid flows through passages formed in the plate itself.

Tube, Structural. See Structural Tube.

Tube-Within-A-Tube. Water-cooled condensing unit in which a small tube is placed inside large unit; refrigerant passes through outer tube, water through the inner tube.

Tubing. Any thin-walled pipe which can be bent easily.

Tubing, Non-Metallic. A round sheath product, of round cross-section, fabricated from a moisture-resistant, flame-retardant material.

Tubular. Having the form of a round or polygonal pipe or tube.

Tubular Lock. See Cylindrical Lock.

Tuck Pointing. The process of removing deteriorated mortar from the surface of an existing brick wall, and inserting fresh mortar.

Tufted Carpet. Carpet made by inserting the pile yarns through a prewoven fabric backing on a machine with hundreds of needles, similar to a huge sewing machine.

Tufts. The cut loops of a carpet pile fabric; applies to both woven and tufted carpets.

Tumbler. A lock mechanism that holds a bolt until operated by a key.

Tumbler Switch. A lever-operated snap switch, powered by electric current

Tumblestone Aggregate. A random collection of aggregate.

Tumbling. Method of finishing or polishing by using a tumbling barrel; articles to be finished and finishing material are put into barrel which is turned or tumbled.

Tung Oil. A yellow drying oil obtained from the seed pods of tung trees and widely used in water-resistant varnishes, lacquers, and high-gloss paints.

Tungsten. 1. A gray-white dense high-melting point ductile hard metallic element that resembles chromium and molybdenum in many of its properties and is used for electric light filaments and in hardening alloys.

Tunnel. A horizontal passageway through or under an obstruction; also called Wolfram. 2. Tungsten Steel.

Tunneling. The act or process of digging a horizontal passageway through or under an obstruction.

Turbidity. Degree to which pool water is visually obscured by suspended particles.

Turbine. A rotary engine actuated by the reaction of a current of fluid, as water, steam, or air, subject to pressure and usually made with a series of curved vanes on a central rotating spindle.

Turf. The layer of grass, matted roots, and earth that comprise the surface of a grassland.

Turn Piece. A small knob, lever, or tee turn with spindle attached for operating the deadbolt of a lock or a mortise bolt.

Turnbuckle. A mechanical device for tightening tie rods by turning a threaded link.

Turner. One who operates a lathe.

Turning Vanes. Thin curved blades placed at bends in air ducts to direct air flow with a minimum of eddying and pressure loss.

Turnkey. A project delivery system where the owner pays an all-inclusive price for the land, financing, design, and construction.

Turnout. 1. A widened space in a highway for vehicles to pass or park. 2. A railroad siding.

Turnover Rate. The number of times a quantity of water, which is equal to the total capacity of water in the swimming pool, passes through the filters in a given period; this is usually expressed in turnovers per day.

Turnstile. A gate for admission or exit with revolving arms allowing people to pass singly.

Turntable. 1. A circular revolving platform for turning a vehicle like a locomotive, truck, bus, or automobile. 2. A lazy susan installed in a corner cabinet.

Turpentine. 1. An oleoresin secreted by various conifers. 2. A volatile pungent oil distilled from turpentine, used in mixing paints and varnishes; also called Turps.

Turps. Turpentine.

Turret. A small tower, often projecting from a building's wall as an ornament.

Turtleback. See Blistering, 2.

Tuscan Red. A red pigment consisting of a combination of iron oxides and a lake.

TV. Television.

TVOC. Total Volatile Organic Compound.

TWA. Time-Weighted Average.

Twist. A distortion caused by the turning or winding of the edges of a board so that the four corners of any face are no longer in the same plane.

Two-Coat System. Two-coat paint application for initial painting.

Two Coat Work. See Double-Up.

Two-Component Gun. One having two separate fluid sources leading to spray head, for spraying a coating and a catalyst simultaneously.

Two-Family Dwelling. A dwelling containing two dwelling units.

Two-Point Latch. A device used to lock an inactive door leaf to the frame head and to the floor.

Two-Pole Motor. 3600 rpm, 60 Hz electric synchronous speed motor.

Two-Temperature Valve. Pressure-opened valve used in suction line on multiple refrigerator installations which maintains evaporators in system at different temperatures.

Two-Way. Construction with steel reinforcing running in two perpendicular directions.

Two-Way Action. Bending of a slab or deck in which bending stresses are approximately equal in the two principal directions of the structure.

Two-Way Concrete Joist System. A reinforced concrete framing system in which columns directly support an orthogonal grid of intersecting joists.

Two-Way Flat Plate. A reinforced concrete framing system in which columns directly support a two-way slab that is planar on both its surfaces.

Two-Way Flat Slab. A reinforced concrete framing system in which columns with mushroom capitals and/or drop panels directly support a two-way slab that is planar on both its surfaces.

Two-Way Valve. Valve with one inlet port and one outlet port.

Tyloses. Masses of parenchyma cells appearing somewhat like froth in the pores of some hardwoods, notably the white oaks and black locust; tyloses are formed by the extension of the cell wall of the living cells surrounding vessels of hardwood.

Tympanum. The recessed triangular face of a pediment.

Typ. Typical.

Type X Gypsum Board. A gypsumboard used where increased fire resistance is required.

Type X Lath. See Gypsum Lath.

Typical. Exhibiting the essential characteristics of a group; a representative example.

U

UBC. Uniform Building Code.

U Bolt. A U-shaped, bent iron bar that has bolts and threads at both ends.

UCC. Uniform Commercial Code.

UF Cable. Underground Feeder Cable.

UHF Cable. Cable that is designed for ultra-high frequency.

UL. Underwriters Laboratories.

ULC. Underwriters Laboratories of Canada.

Ultimate Compressive Strength. The stress at which a material crushes.

Ultimate Load. The absolute maximum magnitude of load which a structure can sustain, limited only by ultimate failure.

Ultimate Strength. Maximum strength that can be developed in a material.

Ultimate-Strength Design. Method of structural analysis of continuous concrete structures based on calculating the loadings which will cause actual failure of sections, rather than the loadings which will cause stresses to reach allowable, safe, values.

Ultramarine. A vivid blue pigment, made from powdered lapiz lazuli.

Ultramarine Blue. Blue pigment made by heating a mixture of China clay, sodium carbonate, carbon and sulphur; it is not affected by alkalis, but is easily affected by acids.

Ultramicroscopic. Too small to be seen with a normal microscope.

Ultrasonic Transmitter. A mechanism that converts vibrations with the same physical nature as sound into equivalent waves above the range of human hearing.

Ultraviolet. 1. The invisible rays of the spectrum at the violet end. 2. Invisible radiation waves with frequencies shorter than wave lengths of visible light and longer than X-ray.

Ultra-Violet Light. Light of short wave length which is invisible but has active chemical effect on finishing materials.

Umber. A natural earth pigment like ochre but darker and browner.

Unadulterated. Not contaminated; pure.

Unbalanced Wall. See Unsymmetrical.

Unbalancing the Schedule of Values. See Front end loading.

Unbonded Construction. 1. Post-tensioned concrete construction in which the tendons are not grouted to the surrounding concrete. 2. A construction contract in which the contractor has not posted a surety bond.

Unbuffed End. An untrimmed, serrated factory cut end.

Uncertainty. The doctrine that contracts are unenforceable if unintelligible

Unconscionability. A doctrine that courts will not enforce contractual provisions that put one party at the mercy of another.

Unctuous. Oily; of minerals, having a greasy or soapy feel.

Uncut loop. Carpet pile yarns that are continuous from tuft to tuft, forming visible loops.

Underatomized. Not dispersed or broken up fine enough.

Underbead Crack. In welding, a crack in the heat-affected zone not extending to the surface of the base metal.

Underbid. To bid less than a competing bidder.

Undercapitalized. Having too little capital for efficient operation.

Undercoat. A coating applied prior to the final or top coat of a paint job; second coat in three-coat work or first coat in repainting.

Undercourse. A course of shingles laid beneath an exposed course of shingles at the lower edge of a wall or roof, in order to provide a waterproof layer behind the joints in the exposed course.

Undercut. A groove at the base metal adjacent to the toe of a weld and left unfilled by weld metal.

Underdrain. 1. Perforated pipe drain installed in crushed stone under a slab to intercept ground water and drain it away from the structure. 2. The distribution system at the bottom of a sand filter to collect the filtered water during a filter run and to distribute the backwash water during backwash.

Underestimate. An estimate that is too low.

Underfloor Duct. A round or rectangular metal pipe placed under a wood floor construction or in a concrete floor to distribute warm air from a heating or air conditioning system.

Underglaze Decoration. A ceramic decoration applied directly on the surface of ceramic ware and subsequently covered with a transparent glaze.

Underground. Subterranean.

Underground Feeder Cable. Insulated electrical cable suitable for underground use.

Underground Paper. Wallpaper without a basic background color.

Underground Tank Removal. The excavation and removal process of a storage tank buried in the ground.

Underground Tank. A container that holds various liquid or solid matter and is found underground.

Underlay. 1. One or more layers of felt applied as required for a base sheet over which finish roofing is applied. 2. See Carpet Padding.

Underlayment. A material placed under finish coverings, such as flooring or shingles, to provide a smooth, even surface for applying the finish.

Underpinning. The process of placing new foundations beneath an existing structure.

Underslab Drainage. The process of continuous interception and removal of ground water from under a concrete slab with the installation of perforated pipe.

Undertone. A color covered up by other colors but when viewed by transmitted light, shows through the other colors modifying the effect.

Underwater Light. A lighting fixture designed to illuminate a pool from beneath the water surface; it can be wet-niche, located in the pool water, or dry-niche, located in the pool wall and serviced from outside the pool.

Undivided Interest. Joint ownership of an interest in the whole property rather than in a specific identifiable part

Unemployment Tax. A program funded by employers based on the wages of employees, paid on a quarterly basis to the Unemployment Insurance Fund from which benefits are paid to qualifying ex-employees.

Unencumbered Property. Real property that has no liens, encumbrances, easements, or debt; free and clear.

Unequal Angle. A steel angle in which the two legs are of different lengths.

Unfaced Insulation. Insulation which has been manufactured without a vapor barrier.

Unfinished Bolt. An ordinary carbon steel bolt.

Unforeseen Conditions. Unexpected or unanticipated events or situations on the jobsite or locality above or below grade that affect the contract price or time.

Unglazed Floor Tile. A hard, dense floor tile deriving color and texture from the materials used and the method of manufacture, with uniform composition throughout.

Unglazed Floor. A hard ceramic floor surface deriving its color or texture from the materials used with uniform composition throughout.

Unglazed Tile. A hard, dense tile of homogeneous composition throughout, deriving color and texture from the materials of which the body is made; the colors and characteristics of the tile are determined by the materials used in the body, the method of manufacture, and the thermal treatment.

Ungrounded Cable. Two-wire nonmetallic sheathed cable that contains one neutral wire and one hot wire.

Uniform Building Code (UBC). A model building code written by the International Building Officials Conference and adopted by many cities and counties as their official building code.

Uniform Sand. A mixture of sand in which most of the granules are the same size.

Union. A type of pipe fitting used to join two pipes in line without turning either pipe.

Union, Brass. A pipe fitting constructed of brass, used to join two pipes without turning either pipe.

Union Hub. A pipe fitting used to join two pipes without turning either pipe, with one end of the fitting enlarged into a bell or socket.

Union, PVC. A pipe fitting used to join two pipes without turning either pipe, manufactured from polyvinyl chloride and used mainly in drain lines because of its resistance to chemicals.

Union, Stainless. A pipe fitting used to join two pipes without turning either pipe, manufactured from stainless steel.

Union T. A pipe tee with a fitting on one end that joins two pipes without turning either pipe.

Unit and Mullion System. A curtain wall system consisting of prefabricated panel units secured with site-applied mullions.

Unitary Package Unit. A system of conditioning air in an outdoor package unit, similar to a rooftop unit, except that the supply and return ductwork passes horizontally through an outside wall.

Unitary System. A heating/cooling system factory assembled in one package and usually designed for conditioning one space or room.

Unit Cost. 1. Cost per unit. 2. Cost per unit of length, volume or area.

Unit Heater. A device for heating a space without the use of ductwork.

Unit Masonry. Manufactured or natural building units of concrete, burned clay, glass, stone, gypsum, etc.

Unit of Work. Calorie, erg, foot-pound, joule, poundal; unit of electrical power, amp, amperage, ampere, kilowatt, megawatt, ohm, voltage, volt, wattage, watt.

Unit Price. 1. The price charged for a single unit or minimum unit of production. 2. The price charged for a unit of length, volume, or area.

Unit Stress. The force on a member divided by its cross-sectional area, such as pounds per square inch.

Unit Substation. A freestanding assembly including a transformer, switchgear and meter.

Unit Vent. A vent pipe which services two or more traps.

Unit Ventilator. A fan coil unit with an opening through an exterior wall to admit outside air.

Unit Water Content. The quantity of water per unit volume of freshly mixed concrete, often expressed as gallons or pounds per cubic yard; this is the quantity of water on which the water- cement ratio is based and does not include water absorbed by the aggregate.

Unit Weight. The weight of a unit volume of a substance; its density.

Universal. Description of a lock, door closer, or other device that can be used on doors of any hand without change.

Universal Motor. Electric motor which will operate on either a-c or d-c.

Unjust Enrichment. A situation in which one party is unjustly enriched at the expense of another party.

Unlawful Detainer Action. A lawsuit brought to evict a defaulting tenant from real property.

Unlined Stack. A chimney built of brick only, possibly lined with a coat of cement plaster, but with no terra cotta flue lining.

Unplasticized Polyvinyl Chloride (uPVC).A stiff plastic material of which doors, windows, corrugated roofing and other building components are made.

Unrated Door. A door which has not been rated for any ability to withstand the spread of fire.

Unreasonable Hardship. This exists when the enforcing agency finds that compliance with the building standard would make the specific work of the project affected by the building standard unfeasible, based on an overall evaluation of all relevant factors, including cost, value, structural stability, safety, conservation, and disabled access.

Unreinforced. Concrete members constructed without steel reinforcing bars or welded wire fabric.

Unsaponifiable Water. Substance in resins and fats which does not unite with caustic alkali to form a soap.

Unstable. A structure that lacks stability.

Unsymmetrical Wall. An assembly of multiple layers of gypsumboard where one side has more layers or a different construction than the other; also called an Unbalanced Wall.

Untreated Joint. A gypsumboard joint without joint compound, tape, or battens.

Uplift. An upward force.

Upset Price. A guaranteed maximum price in a construction contract.

Upside Down Roof. A membrane roof assembly in which the thermal insulation lies above the membrane.

uPVC. Unplasticized Polyvinyl Chloride.

Urban Renewal. The redevelopment of a deteriorated inner city area.

Urea. A crystalline colorless nitrogenous compound found primarily in the urine of mammalian animals; can be synthesized from carbon dioxide and ammonia for use in fertilizers, resins, and plastics.

Urea-Formaldehyde Foam. An aerated foam of a thermosetting synthetic resin composed of urea and formaldehyde; used for thermal insulation.

Urea-Formaldehyde Resin. Product obtained by chemical reaction between urea and formaldehyde in presence of catalyst.

Urea-Melamine Resin. Melamine-formaldehyde resins which produce tough finish that approaches porcelain.

Urethane Board. Rigid form of plastic foam of polyurethane.

Urethane Foam. Type of insulation which is foamed in between inner and outer walls of a container.

Urethane Insulation. Board form or sprayed on plastic foam of rigid polyurethane.

Urethane Padding. Plastic foam of polyurethane used under carpeting for cushioning.

Urethane Resins. A particular group of film formers; Isocyanate resins.

Urethane, Rigid. Sheet or board form of plastic foam of rigid polyurethane.

Urethane Roof. A roof with plastic foam of rigid polyurethane; made into board form or sprayed on.

Urinal. A water flushed plumbing fixture designed to receive urine directly from the user of the fixture.

Urinal Carrier. A metal structure attached to a bathroom wall to support and mount a urinal.

Urinal Screen. A privacy panel that separates urinals from each other; the screens are either floor, wall or ceiling mounted.

Urn, Coffee. See Coffee Urn.

Useful Life. The length of time a coating is expected to remain in service.

USP. Letters affixed to name of material to indicate that it conforms in grade to specifications of United States Pharmacopoeia and that it is approved for use in medicinal preparations. The material does not necessarily have to be chemically pure.

U Stirrup. An open-top, U-shaped loop of steel bar used as reinforcing against diagonal tension in a beam.

Usury. Charging more than the maximum legally allowable rate of interest.

Utensil Washer, Medical. See Medical Utensil Washer.

Utility. A service provided by a public utility, such as electrical power or water.

Utility Excavation. The act or process of either digging up existing cable buried in the ground, or trenching to lay new cable.

Utility Knife. 1. A commonly used razor-type blade and holder. 2. See Board Knife.

Utility Pole. A vertical pole with cross arms, where electrical and telephone lines are carried.

Utility Room. A room housing domestic appliances such as a laundry.

Utilization Equipment. Equipment which utilizes electric energy for mechanical, chemical, heating, lighting, or similar useful purposes.

UV. Ultraviolet.

U-Value. See Coefficient of Heat Transmission.

V

V. A symbol for vertical shear.

VA. Veterans Administration.

Vacations. The uncoated portion of a finished object; also called Skips and Holidays.

Vacuum. 1. A space absolutely devoid of matter. 2. Pressure lower than atmospheric pressure.

Vacuum Breaker. 1. An electrical breaker with a space that contains reduced air pressure. 2. A device which prevents the formation of a vacuum in a water supply pipe; installed to prevent backflow.

Vacuum Cleaner. One of several types of suction devices used to collect dirt from the bottom of a swimming pool; these units work in a variety of ways - some force collected dirt to the filter, others to the waste drain, or into a porous container.

Vacuum Control System. Intake manifold vacuum is used to operate dampers and controls in some automobile systems.

Vacuum Gauge. Instrument used to measure pressures below atmospheric pressure.

Vacuum Piping. The pipe from the suction side of a pump connected to a vacuum fitting located at the pool and below the water level to which underwater cleaning equipment may be attached.

Vacuum Pump. Special high efficiency device used for creating high vacuums for testing or drying purposes.

Valley. 1. An elongated depression in the earth's surface between two raised areas. 2. A trough or internal angle formed by the intersection of two roof slopes.

Valley Flashing. Thin sheet metal used to line the valley of a roof.

Valley Rafter. A diagonal rafter that supports a roof valley.

Valuation. The opinion of a real estate appraiser as to the value of a parcel of real property.

Value. 1. The worth of something measured monetarily. 2. Term used to distinguish dark colors from light ones; dark values are known as shades; light values as tints.

Value Engineering. The process of analyzing the cost versus the value and alternative materials, equipment, and systems.

Value of the Assets. The assets of a company in terms of dollar value.

Valve. Numerous mechanical devices by which the flow of liquid, gas, or loose material in bulk may be started, stopped, or regulated by a movable part that opens, shuts, or partially obstructs one or more passageways.

Valve, Expansion. See Expansion Valve.

Valve Plate. Part of a compressor located between the top of the compressor body and the head, containing compressor valves and ports.

Valve, Solenoid. See Solenoid Valve.

Valve, Suction. See Suction Valve.

Valve, Tapping. See Tapping Valve.

Valve, Water. See Water Valve.

Van Dyke Brown. A brown pigment which consists of decomposed vegetable matter that has almost approached the coal state; it is weak in hiding power compared to umber and sienna.

Vanadium. 1. A hard gray malleable ductile metallic element found in several ores and used as an alloy in small quantities for strengthening some steels. 2. Vanadium steel.

Vane. A revolving pointer mounted on a high place to show the direction of the wind.

Vanity Cabinet. Case, box, or piece of furniture which rests on the floor and receives a lavatory, commonly has shelves and doors and is primarily used as storage for below lavatory.

Vanity, Bath. See Bath Vanity.

Vapor (Vapour). The gaseous state of any substance.

Vapor Barrier. 1. A material with a high resistance to vapor movement, such as foil, plastic film, or specially coated paper, that is used in combination with insulation to control condensation. 2. Any material that has a water vapor permeance (perm) rating of one or less. 3. A type of plastic sheeting that both eliminates drafts and keeps moisture from damaging a building or structure. 4. A waterproof membrane placed under concrete floor slabs that are on grade. 5. Material used to retard the movement of water vapor into walls, and prevent condensation in them; applied separately over the warm side of exposed walls or as a part of batt or blanket insulation. 6. Thin plastic or metal foil sheet used in air-conditioned structures to prevent water vapor from penetrating insulating material.

Vapor Compression Refrigeration. A cooling process that transforms heat from one location to another; heat is forced to flow from a region of lower temperature to a region of higher temperature; similar to a refrigerator for cooling food.

Vapor Degreasing. A cleaning process utilizing condensing solvent as the cleaning agent.

Vaporization. Conversion from liquid or solid to a gaseous state.

Vapor Lock. Condition where liquid is trapped in line because of bend or improper installation; such vapor prevents liquid flow.

Vapor Migration. The movement of water vapor from a region of high vapor pressure to a region of lower vapor pressure.

Vapor Pressure. Pressure imposed by either a vapor or gas.

Vapor Pressure Curve. Graphic presentation of various pressures produced by refrigerant under various temperatures.

Vapor Retarder. A layer of material intended to obstruct the passage of water vapor through a building assembly; also called a Vapor Barrier; sealing material placed between building components or materials to arrest movement of moisture within building enclosure assembly.

Vapor Retarder Felt. Sealing material made of asphalt-saturated felt.

Vapor, Saturated. See Saturated Vapor.

Variable. Changing; subject to change, not constant; indeterminate.

Variable Air Volume (VAV). An air handling system, unlike a single zone, terminal reheat, multi-zone, or dual duct, supplies air at a constant temperature and varies the air quantity supplied to each zone.

Variable Pitch Pulley. Pulley which can be adjusted to provide different pulley drive ratios.

Variable Rate Mortgage. A mortgage that allows the lender to raise or lower the interest rate from time to time as some specified index rises or falls.

Variance. A deviation in a building or zoning ordinance granted by an appeal authority upon relevant grounds being proven.

Varnish. A thickened preparation of drying oil or drying oil and resin; when applied to a surface, it leaves a hard, glossy, transparent coating; may also be mixed with pigments to make enamels; clear varnish is a slightly yellow, semitransparent liquid.

Varnishing. Application of varnish to a surface using a brush.

Varnish-Stain. Interior varnish tinted with pigments or dyes.

Vat Dyes. Dyes formed in fabrics by oxidation and precipitation of the original dye liquor, for example, indigo; vat dyeing refers to a kind of dye rather than a method of dyeing; raw stock dyeing, skein dyeing, or solution dyeing can be performed with vat dyes.

Vault. 1. An arched surface. 2. An arch translated along an axis normal to the plane of its centerline curve. 3. A room to store valuable items.

Vault Door. A hinged, pivoted, or sliding member that permits passage to an enclosure built for safety or security.

VAV. Variable Air Volume.

V Beam Roof. A roof with corrugated sheeting with flat, V-angled surfaces.

V-Belt. Type of belt commonly used in refrigeration work; it has a contact surface with the pulley which is in the shape of the letter V.

VCP. Vitreous Clay Pipe.

VDU. Video Display Unit.

Vector. 1. A quantity having direction as well as magnitude; usually represented by a line in space with an arrowhead indicating direction and the scale of the line representing magnitude. 2. A carrier of disease, such as a roach, rat, or mouse.

Vegetable Oils. Oils obtained from the seeds or nuts of vegetable growth, including linseed, soybean, perilla, tung, or castor.

Vegetation Control. The act or process of spraying chemicals or placing powders on unwanted vegetation.

Vehicle. 1. Any conveyance for transporting people or goods. 2. The liquid portion of paint and other finishing materials, consisting of the binder (nonvolatile) and thinners (volatile).

Vehicle Lift. 1. An apparatus for lifting vehicles in order to have access to the carriage underneath. 2. A mechanical device used to move vehicles up and down in a parking structure.

Vehicular Way. A route used for vehicular traffic, such as a street, driveway or parking lot.

Veining. The characteristic stretch marks that develop during the aging process of soft bitumen.

Vellum. Fine parchment-like paper used for drawings for reproduction.

Vellum Glaze. A semi-mat glaze having a satin-like appearance.

Velocimeter. Instrument used to measure air speeds using a direct-reading air speed indicating scale.

Velocity. Speed, usually expressed in length units per time unit, like miles per hour or feet per second.

Velocity of Sound. A product of frequency times wavelength.

Velvet Carpet. A pile fabric woven on a velvet loom; it is the simplest of all carpet weaves and is used mostly for solid colors; woven with wires, a looped pile is created when the wires are withdrawn; a cut pile results when knife blades on the ends of the wires cut the loops; tightly twisted yarns in cut pile provide a frieze surface; varied heights of looped pile can be achieved by using shaped wires; see Woven Carpet

Veneer. 1. A thin layer, sheet or facing. 2. A thin layer or sheet of wood; usually one that has beauty or value and is intended to be overlaid on an inferior surface. 3. Nonstructural facing of brick, concrete, stone, tile, metal, plastic or other similar approved material attached to a backing for the purpose of ornamentation, protection or insulation.

Veneer, Ashlar. See Ashlar Veneer.

Veneer, Brick. See Brick Veneer.

Veneer, Granite. See Granite Veneer.

Veneer Plaster. A wall finish system in which a thin layer of plaster is applied over a special gypsumboard base.

Veneer Plaster Base. A gypsumboard used as the base for application of a gypsum veneer plaster.

Venetian Blind. A window blind which has numerous horizontal slats that may be set at several different angles in order to vary the amount of light admitted.

Venetian Red. Pigment with brick-red color made synthetically by calcining copperas and whiting.

Venetian Terrazzo. A mosaic type of terrazzo topping in which large chips of stone are incorporated.

Vent. 1. A vertical pipe connected to a waste or soil distribution system that prevents a back pressure or a vacuum that might siphon the water out of a trap. 2. Vertical pipe to provide passageway for expulsion of water vapor and vent gases from gas-burning equipment to outside air. 3. A free opening to provide air intake, expulsion, or circulation in such areas as underfloor crawl spaces and attics.

Vent Cap. The top piece installed over the terminus of a ventilation pipe.

Vent, Foundation. See Foundation Vent.

Ventilated. Provided with a means to permit circulation of air sufficient to remove an excess of heat, fumes or vapors.

Ventilation Air. 1. The total volume of air in a system, which is a combination of the air brought into the system from the outdoors and the air that is being recirculated within the building. 2. The fresh air brought into the system from the outdoors.

Ventilator, Gravity. See Gravity Ventilator.

Ventilator, Relief. See Relief Ventilator.

Venting Clip. Clips to provide means to aid in curing and exiting of moisture from concrete or gypsum fill placed on a deck.

Venting of Roof Assembly. The necessary venting to eliminate moisture and vapor in roof systems.

Vent, Masonry. See Masonry Vent.

Vent, Roof. See Roof Vent.

Vent Screed. A sheet metal plaster screed incorporating a vented section used chiefly in roof overhang soffits.

Vent, Smoke. See Smoke Vent.

Vent, Soffit. See Soffit Vent.

Vent Stack. A plumbing vent pipe in a multistory building, a separate pipe used for venting, that either connects with a stack vent above the highest fixture, or extends through the roof.

Venturi. A short tube with a tapering constriction in the middle that causes an increase in the velocity of flow of a fluid and a corresponding decrease in fluid pressure.

Verdigris. 1. A green or bluish-green crystallized substance formed on copper by the action of acetic acid, used as a pigment. 2. Green oxidation on copper, bronze, or brass.

Verge Boards. The boards which serve as the eaves finish on the gable end of a building.

Vermiculite. A mineral closely related to mica, with the faculty of expanding on heating to form lightweight material with insulation quality; used as bulk insulation and also as aggregate in insulating and acoustical plaster and in insulating concrete floors.

Vermiculite Insulation. A loose mineral fill for thermal insulating applications.

Vermiculite Plaster. 1. A covering that is put on steel beams, concrete slabs, and other heavy construction materials, as a fire-retardant. 2. An insulating and soundproofing plaster.

Vermilion. Sulphide of mercury used as a pigment.

Vermont Slate. A fine grained thin-layered rock used for roofing, and paving.

Vernacular. The common building style of a period or place; concerned with ordinary rather than monumental buildings; vernacular architecture.

Vernier. A supplementary scale used on an instrument for obtaining fine adjustment.

Vertical. Perpendicular to the plane of the horizon; upright.

Vertical Angle. An angle in a vertical plane.

Vertical Application of Roofing. Roll roofing laid parallel to the roof slope.

Vertical Application. Of gypsumboard, see Parallel Application.

Vertical Bar. An upright reinforcing bar in a reinforced concrete shape.

Vertical Broken Joint. Style of laying tile with each vertical row of tile offset for half its length.

Vertical Check Valve. A device, mounted vertically, which allows fluid or air to pass through in only one direction.

Vertical Curve. The meeting of different gradients in a road or pipe.

Vertical Drains. Usually columns of sand used to vent water squeezed out of humus by weight of fill.

Vertical Forces. Loads imposed on a structure caused by gravity or the vertical component of wind or earthquake forces.

Vertical-Grained Wood. Edge-grained lumber.

Vertical Joint. See Butt Joint, 4.

Vertical Lift Door. A door assembly on a lifting or hoisting device that is opened or closed, bottom to top or top to bottom.

Vertical Louver. A louver equipped with slats that are mounted vertically; see Louver.

Vertically Laminated Wood. Laminated wood in which the laminations are so arranged that the wider dimension of each lamination is approximately parallel to the direction of load.

Vertical Pattern. A spray pattern whose longest dimension is vertical.

Vertical Pipe. Any pipe or fitting which makes an angle at 45 degrees or less with the vertical.

Vertical Position. In pipe welding, the position of a pipe joint wherein welding is performed in the horizontal position and the pipe may or may not be rotated.

Vertical Reinforcing. Steel reinforcing bars in concrete, running vertically, perpendicular to the horizontal bars.

Vertical Shear. The sum of the algebraic forces that are one side of a given cross-section of a member.

Vertical Siding. Exterior wall covering attached vertically to the wood frame of a building or structure.

Vertical Slotted Shelf Standard. Metal strip with slots attached to vertical surfaces as the support part of adjustable shelf hardware.

Vertical Transportation. A general term for elevators, lifts, escalators, and dumbwaiters.

Vertical Wood Board and Batten Siding. Linear vertical wood boards with wood strips covering vertical joints used as an exterior cladding for a framed wall.

Vertical Wood Board Siding. Linear vertical wood material used as exterior surface or cladding for exterior framed wall.

Vessels. Wood cells of comparatively large diameter that have open ends and are set one above the other to form continuous tubes; the openings of the vessels on the surface of a piece of wood are usually referred to as pores.

Vestibule. An entrance to a house; usually enclosed.

VG. Vertical Grain.

Vibration. 1. A periodic motion which repeats itself after a definite interval of time. 2. Energetic agitation of concrete to assist in its consolidation, produced by mechanical oscillating devices at moderately high frequencies.

Vibration Aresstors. Soft or flexible substance or device which will reduce the transmission of a vibration.

Vibration Detector. A part of a burglar alarm system.

Vibration Isolation. A method of isolating equipment from a structure to reduce the vibration transmission from the equipment, thus reducing the sound transmitted to the structure.

Vibration Isolator. A flexible device supporting vibrating mechanisms or machinery and reducing the vibrations transmitted to the rest of the building or structure.

Vibrator. Device for agitating wet concrete to eliminate air pockets.

Vibrator, External. See External Vibrator.

Vibrator, Internal. See Internal Vibrator.

Vicat Apparatus. A penetration device used to determine the setting characteristics of hydraulic cements.

Video Display Unit (VDU). 1. Cathode Ray Tube. 2. A television screen. 3. A computer monitor.

Vierendeel Truss. A rigid frame, used as a beam, assembled from parallel top and bottom chords tied together by vertical members.

Vinyl. A plastic made by polymerization, used in flooring, wall coverings, gaskets, and miscellaneous parts.

Vinyl Asbestos Tile. Resilient floor tile consisting of vinyl reinforced with asbestos fibers.

Vinyl Coated Fence. A fence or enclosure built with posts and rails and covered with chain-link which has been factory coated with plastic.

Vinyl Coating. One in which the major portion of the binder is of a vinyl resin.

Vinyl Copolymer. Resins produced by copolymerizing vinyl monomers such as vinyl acetate and vinyl chloride.

Vinyl-Covered Gypsum Board. See Predecorated Wallboard.

Vinyl Faced Panel. Sheathing that has been covered by a flexible film or a liquid bonded by heat.

Vinyl Floor Tile. A resilient floor tile made of vinyl.

Vinyl Foam Cushioning. Carpet cushioning made from a combination of foamed synthetic materials.

Vinyl Gasket. A plastic seal between glass and aluminum.

Vinyl Resins. Synthetic resins made from vinyl compounds such as vinyl acetate.

Vinyl Sheet Floor. A thin sheet of vinyl used for the finish of floor surfaces.

Vinyl Sheet. The rolled form of vinyl.

Vinyl Sheetrock. Gypsumboard with a thin layer of vinyl as the finished surface.

Vinyl Siding. Exterior wall coverings made from a thermoplastic compound.

Vinyl Tile. A semi-flexible, resilient floor tile made from polymerized vinyl chloride, vinylide chloride, or vinyl acetate.

Vinyl Trim. Extruded vinyl moldings used for concealing edges, ends, joints, and corners of various sheet materials.

Vinyl Wall Covering. A tough, flexible and shiny film or liquid bonded by heat to a paper or fabric backing material.

Virgin Growth. The growth of mature trees in the original forests; to be distinguished from Second Growth.

Viscoelasticity. The ability of a material to simultaneously exhibit viscous and elastic responses to deformation.

Viscometer. An instrument for the measurement of Viscosity.

Viscosity. Internal friction in a fluid; resistance to flow; the ratio of the shear stress existing between laminae of moving fluid and the rate of shear between these laminae.

Viscosity Cup. A device for measuring viscosity.

Viscosity Grading. A classification system of asphalt cements based on viscosity ranges at 140°F (60°C); a minimum viscosity at 275°F (135°C) is also usually specified; the purpose is to prescribe limiting values of consistency at these two temperatures; 140°| F approximates the maximum temperature of asphalt pavement surface in service in the U.S.; 275?F approximates the mixing and laydown temperatures for hot asphalt pavements; there are five grades of asphalt cement based on the viscosity of the original asphalt at 140?F.

Vise. Bench tool with two jaws that can be closed by turning a screw, for holding work.

Vise Grip. A pliers-like hand tool that has great gripping power through a simple system of highly efficient leverage.

Vision Panel. Glass placed in an opening of a door.

Visual Aid Board. A display that has information posted on it with writing, diagrams or pictures to dispense helpful facts.

Vitreous. That degree of vitrification evidenced by low water absorption; vitreous generally signifies less than 0.5% absorption, except for floor and wall tile and low-voltage electrical porcelain which are considered vitreous up to 3.0% water absorption.

Vitreous China. The material of a china plumbing fixture with a finish resembling glass, in color, composition, brittleness, and low porosity, used for toilets, urinals, and lavatories.

Vitreous Enamel. A fired-on opaque glassy coating on metal, such as steel.

Vitreous Slip. A slip coating matured on a ceramic body, producing a vitrified surface.

Vitreous Tile. Tile with water absorption of more than 0.5%, but not more than 3.0%.

Vitrification. The progressive reduction in porosity of a ceramic composition as a result of heat treatment, or the process involved.

Vitrification Range. The maturing range of a vitreous body, producing a vitrified surface.

Vitrified. That which is fused by heat; well burned to a greater degree of hardness.

Vitrified Clay Pipe. Pipes used especially for underground drainage, that are made of hard baked clay.

VM&P Naphtha. Varnish and paint manufacturers naphtha; a low power flammable hydrocarbon solvent.

VOCs. Volatile Organic Compounds.

Void. 1. An unfilled space in a material, of trapped air or other gas. 2. In gypsumboard, a hollow space in the core caused by the entrapment of air during the manufacturing process. 3. In concrete, air spaces resulting from segregation and improper consolidation during placement. 4. In asphalt paving, empty spaces in a compacted mix surrounded by asphalt coated particles

Voidable. A valid provision of a contract that can be made void by following some specified procedure.

Void Volume. Total empty spaces in a compacted mix.

Volatile Content. Those materials which evaporate; usually expressed as a percentage.

Volatile Flammable Content. A flammable liquid having a flash point below 100°F. (37.8°C.) or a liquid whose temperature is above its flash pint.

Volatile Organic Compounds (VOCs). Compounds that evaporate from the many housekeeping, maintenance, and building products made with organic chemicals; these compounds are released from products that are being used and that are in storage; in sufficient quantities, VOCs can cause eye, nose, and throat irritations, headaches, dizziness, visual disorders, memory impairment; some are known to cause cancer in animals; some are suspected of causing, or are known to cause, cancer in humans; at present, not much is known about what health effects occur at the levels of VOCs typically found in public and commercial buildings.

Volatiles. Fluids which evaporate rapidly.

Volatile Thinner. A liquid that evaporates readily and is used to thin or reduce the consistency of finishes without altering the relative volumes of pigments and nonvolatile vehicles.

Volcanic Ash. A fine powder similar to diatomite but lighter in weight, it is used as a filter medium or filter aid in diatomite filters.

Volt. A unit of electrical potential difference; the SI unit of electromotive force; the difference of potential that would carry one ampere of current against one ohm resistance.

Voltage. 1. Term used to indicate the electrical potential or electromotive force in an electrical circuit. 2. Voltage or electrical pressure which causes current to flow. 3. Electromotive force.

Voltage Control. Device used to provide some electrical circuits with uniform or constant voltage.

Voltage Drop. Loss of voltage due to length of run or resistance.

Voltage Regulator. An automatic electrical control device for maintaining a constant voltage supply to the primary of a welding transformer.

Voltage to Ground. In grounded circuits the voltage between the given conductor and that point or conductor of the circuit which is grounded; in ungrounded circuits, the greatest voltage between the given conductor and any other conductor of the circuit.

Voltmeter. An instrument for measuring in volts the differences of potential between different points of an electrical circuit.

Volume. The amount of space occupied by a three dimensional solid or gas, measured in cubic units like cubic feet or cubic centimeters.

Volume Ceiling. Any ceiling higher than the normal.

Volumetric Efficiency. The relationship between the actual performance of a compressor or of a vacuum pump and calculated performance of the pump based on its displacement.

Volute. 1. A spiral scroll characteristic of Ionic capitals, also used in Corinthian and composite capitals. 2. A spiral or scroll-shaped form, as in volute pumps.

Vomitory. One of the passages for entrance and exit in a theater or amphitheater.

Vortex Tube Refrigeration. Refrigerating or cooling device using principle of vortex tube, as in mining suits.

Vortex Tube. Mechanism for cooling or refrigerating which accomplishes cooling effect by releasing compressed air through a specially designed tube.

Voussoir. One of the wedge-shaped pieces of stone or brick forming an arch.

VRM. Variable Rate Mortgage.

VTR. Vent through roof.

Vulcanizing. Treating rubber with sulphur at a high temperature to harden and toughen it.

W

W. The total dead load used in earthquake design.

W/C Ratio. Water Cement Ratio.

Wadding. The act of hanging staff by fastening wads made of Plaster of Paris and excelsior or fiber to the casts and winding them around the framing.

Wading Pool. Any constructed or prefabricated pool used for wading which is less than twenty-four (24) inches in depth.

Waferboard. A building panel made by bonding together large, flat flakes of wood.

Waferboard Sheathing. Exterior wall or roof covering panels of waferboard

Waffle Slab. 1. A two-way concrete joist system. 2. Two-way slab or flat slab made up of a double system of narrow ribs or joists, usually at right angles to each other, forming a pattern of waffle-like coffers.

Wagner Fineness. The fineness of materials such as portland cement expressed as total surface area in centimeters per gram as determined by the Wagner turbidimeter apparatus and procedure.

Wainscot. The lower 3 or 4 feet of an interior wall when it is finished differently from the remainder of the wall.

Wainscot, Terrazzo. See Terrazzo Wainscot.

Waiver. The intentional relinquishment of a known right.

Wale. A horizontal beam.

Waler. A horizontal structural member in a concrete forming or excavation shoring system.

Walk. A path specially arranged or paved for walking.

Walk-In Cooler. Larger commercially refrigerated space kept below room temperature; usually installed in supermarkets, restaurants, food processing plants, and wholesale meat distribution centers.

Walk-Up. A two or more story apartment building or living unit without an elevator.

Walkway, Roof. See Roof Walkway.

Wall. A member, usually vertical, used to enclose or separate spaces.

Wall Angle. 1. An L-shaped aluminum angle used as the wall termination of T-bar suspended ceiling systems to support the perimeter acoustic tiles. 2. A structural steel angle attached to a masonry wall.

Wall Blocking. Framing lumber cut in short lengths and installed horizontally between wall studs as filler pieces to stabilize the framing, to act as fire blocking, or to provide a backing for fastening a finish item.

Wallboard. Large, rigid sheets of wood pulp, gypsum, pressed cellulose fibers, gypsumboard, plywood, or similar materials used in place of plaster in interior surfaces.

Wall Bracket. 1. A wall-mounted support for shelving or other object. 2. A wall-mounted lighting fixture.

Wall Cabinet. Case, box, or piece of furniture which mounts on a wall, commonly with shelves and doors, used for storage.

Wall, Cavity. See Cavity Wall.

Wall Cladding. Exterior wood or metal building siding.

Wall Cleanout. An opening in a wall for removal of refuse.

Wall Covering. Any of a variety of final applications to finish a wall surface.

Wall Expansion Joint. A break or space in wall construction to allow for thermal expansion and contraction of materials.

Wall Fabric. Wallpaper or other fabric used to cover or finish off an existing wall.

Wall Finish. The final planing, sanding, staining, varnishing, waxing, or painting of a wall.

Wall Flange. A ridge on a wall that prevents movement. A supporting rim on a wall for attachments.

Wall Footing. A continuous spread footing that supports a uniform load from a wall.

Wall Formwork. The system of wood support for a freshly placed concrete wall left in place until the concrete has set.

Wall Framing. Building construction where exterior and other bearing walls are made of wood.

Wall Furring. Strips of wood applied to make a wall surface level, form an air space, or provide fastening surfaces.

Wall Grille. A wall grating used to cover an opening as protection or as an ornament.

Wall Heater. A heating unit installed in or on a wall.

Wall-Hung. Anything hung from a wall, such as cabinets or plumbing fixtures.

Wall Hung Sink. A lavatory mounted on brackets attached to a wall.

Wall Hydrant. A connection to a water main cut through and mounted on a wall; see Hose Bibb.

Wall Insulation. Material placed in wall cavities for the reduction of fire hazard or for protection from heat and cold.

Wall Lath. Expanded metal, gypsum sheets, or thin strips of wood attached to stud walls, acting as a foundation for plastering.

Wall Louver. Openings in walls for ventilating spaces.

Wall Mounted Flagpole. A pole to raise or display a flag, mounted or attached to a wall.

Wall Mounted Heater. A heating unit mounted on or attached to a wall.

Wall Mounted Oven. A domestic oven for cooking purposes designed for mounting in or on a wall or other surface.

Wall Panel. Form sheathing, constructed from plywood, boards, or metal sheets, that are installed as a unit.

Wallpaper. A decorative paper for applying to the walls of a room.

Wallpaper Paste. An adhesive used for attaching wallpaper.

Wall Pier. A wall segment with a horizontal length to thickness ratio between 2.5 and 6 and whose clear height is at least two times its horizontal length.

Wall Plaster. A paste-like composition that hardens on drying and is used for coating walls.

Wall Railing. A band of ornamental wood installed horizontally.

Wall Reinforcing, Masonry. Steel reinforcing rods or mesh used in masonry walls between courses.

Wall Reinforcing. To strengthen a wall by the addition of new or extra materials.

Wall, Retaining. See Retaining Wall.

Wall Scupper. See Scupper, 2.

Wall Sheathing. The first layer of covering on an exterior wall, fastened to the wall studs.

Wall Size. Solution such as glue, starch, casein, shellac, varnish or lacquer, used to seal or fill pores of wall surface to stop suction, counteract chemicals or stains and prepare surface for paint, paper or fabric.

Wall Spreader. An accessory, usually fabricated from reinforcing bar to a Z or U shape, used to separate and hold apart two faces of curtains of steel reinforcements in a concrete wall.

Wall System. A modular set of shelves and cabinets that can be arranged along a wall in various combinations.

Wall Tie. A mechanical metal fastener which connects wythes of masonry to each other or to other materials.

Wall Tile. A glazed tile with a body that is suitable for interior use and which is usually non-vitreous, and is not required nor expected to withstand excessive impact.

Wall-to-Wall. Covering the entire floor area.

Wall Z Tie. A Z-shaped reinforcing strip used as a support bracket from the structural wall to the masonry veneer.

Walnut. The richly grained moderate reddish-brown wood of the walnut tree, used for veneers, cabinetmaking, and moldings.

Walnut Door. A door which has a veneer of walnut.

Walnut Veneer. An overlay of a thin layer of walnut wood for outer finish or decoration.

Wandering Block Sequence. A block sequence wherein successive blocks are completed at random after several starting blocks have been completed.

Wane. Bark or lack of wood from any cause on edge or corner of a piece of lumber except for eased edges.

Ward. 1. A large room in a hospital, that will accommodate several patient beds. 2. The inner court of a castle or fortress. 3. A local Mormon congregation. 4. A projecting ridge of metal on a lock to prevent insertion of a key that does not have a matching notch.

Wardrobe. A room or freestanding closet where clothes are kept.

Wardrobe Locker. A clothes storage cabinet with a locking door.

Warehouse. A storage room or building.

Warm Air Heating. Any heating system which depends upon the circulation of warm air.

Warm Colors. Colors in which red-orange predominates; so termed because of the association with fire, heat, and sunshine.

Warming Colors. Any color except green may be warmed by adding red; green is warmed by adding yellow.

Warp. 1. Any variation from a true or plane surface. 2. Warp includes bow, crook, cup, and twist, or any combination thereof. 3. In carpet, the backing yarns running lengthwise.

Warpage. A concave or convex curvature of a material that was intended to be perfectly flat.

Warranty. A promise that a fact is true.

Warren Truss. A truss in which the top and bottom chords are parallel or nearly so, the top chord sometimes sloping for roof drainage.

Wash. The slant of a sill, wall, parapet, or capping, to allow the water to run off easily.

Washboarding. In asphalt paving, see Corrugations.

Wash Coat. A very thin coat of finishing material, usually shellac.

Washer. A flat thin ring or a perforated plate used in joints or assemblies to ensure tightness or relieve friction.

Washer, Coin. See Coin Washer.

Washer-Extractor. A clothes washer that includes a high speed centrifugal drying cycle that removes all the free water except for dampness.

Washer, Flat. See Flat Washer.

Washer, Steel Welding. See Steel Welding Washer.

Wash Fountain. A waist high sink which supplies a steady stream of water to cleanse the hands.

Washing. 1. Rapid dissolution or emulsification of a paint film when wet with water. 2. Erosion of a paint film after rapid chalking.

Wash-Out. Lack of proper coverage and texture build-up in machine-dash textured plaster caused by the mortar being too soupy.

Wash Primer. A thin inhibiting paint usually chromate pigmented with a polyvinyl butyrate binder.

Washroom Faucet. A device that dispenses hot and cold water, mounted above a sink.

Waste. 1. Valueless material. 2. Digging, hauling, and dumping of valueless material to get it out of the way. 3. Liquid discharged from any plumbing fixture, except water closets and urinals.

Waste and Overflow Fitting. A bathtub drain fitting which provides both the outlet for the bathtub drain and an overflow to drain excess water from the tub.

Waste Handling. The act or process of transporting or removing rubbish.

Waste Pipe. A plumbing line that carries waste water from fixtures except water closets and urinals.

Waste Piping. See Filter Waste Discharge Piping.

Waste Receptacle. A container for the temporary storage of rubbish.

Waster Mold. An unreusable precast plaster mold made for the forming of decorative monolithic or cast-in-place concrete; the mold cannot be removed without being destroyed.

Waste Stack. A vertical line of piping that extends one or more floors and receives the discharge of fixtures other than water closets and urinals.

Waste Water. In a swimming pool, the water from any filter, perimeter overflow, pool emptying line, or similar apparatus or appurtence.

Watchman. A guard who keeps watch over a certain area.

Water. A colorless, transparent, odorless compound of hydrogen and oxygen, H_2O.

Water Absorption. 1. The amount of water absorbed by a material under specified test conditions; expressed as weight percent of the test specimen. 2. The amount of water ingested into the core and surface papers of gypsumboard; expressed as a percent of water added over dry weight.

Waterbar. A rubber or plastic strip cast into concrete at joints to prevent water leakage; also called Waterstop.

Water Blasting. Blast cleaning using high velocity water.

Water-Cement Ratio. The proportion of water to portland cement; the number of gallons of water per 94 pound sack of cement; only a small amount of water is needed to hydrate the cement and complete the chemical reaction; all additional water is for workability only and too much of which will weaken the concrete or mortar.

Water Closet. See Toilet.

Water Conditioner. A device used to dissolve minerals from water; normally after the minerals are removed, the water tastes better and reduces the likelihood that mineral deposits will build up within the plumbing system; the water is considered to be soft after the minerals have been removed.

Water Cooled Chiller. A piece of equipment that produces chilled water for circulation through a building and used for cooling.

Water Cooled Condenser. A condenser in which water is normally circulated through a cooling tower through which heat is dispersed to the atmosphere.

Water Cooler. An apparatus that cools, holds, and dispenses cold water.

Water Defrosting. Use of water to melt ice and frost from evaporator during off-cycle.

Water Gain. Of concrete, see Bleeding, 2.

Water Gauge. An instrument to measure the depth of water, or to indicate the height of its surface, as in a steam boiler.

Water Glass. 1. An instrument consisting of an open box or tube with a glass bottom, used for examining objects in or under water. 2. A Water Gauge. 3. A viscous, syrupy solution of sodium or potassium silicate that is used especially as a cement, as a protective coating and fireproofing agent, and in preserving eggs.

Water Hammer. A banging sound in water supply lines caused by the sudden stopping of the water flow.

Water Heater. An apparatus for heating and storing water.

Water Level.. A length of clear plastic hose, 3/8 to 1/2 inch in diameter and approximately 50 feet long, filled with water and used as a leveling device to check level in walls or structures.

Water Loss. A measurement of the amount of free water evaporated from gypsumboard products during the drying stage in the manufacturing process; expressed in pounds per 1000 S.F.

Water Main. Water supply pipe generally located at the street which may supply a number of buildings.

Water Meter. An instrument for measuring water consumption.

Water of Hydration. The water required to replace the water lost during the calcination process.

Water Paint. A paint in which the vehicle is a water emulsion and in which water is used as thinner.

Water Pipe Restoration. To install new linings in pipes to improve the condition of existing water pipes; usually includes the cleaning and removal of built-up scale or debris out of the existing pipe and the relining with a compatible material.

Water Piping Removal. The act or process of tearing out and carrying away old water piping systems.

Waterproof. So constructed or protected that moisture will not interfere with its successful operation.

Waterproof Cement. Portland cement to which waterproofing agents, such as surface repellents, have been added at time of blending materials at the mill.

Waterproofing. 1. The act or process of making something waterproof. 2. A coating capable of stopping penetration of water or moisture.

Waterproof Membrane. An impermeable layer of plastic film, roofing material, bitumen, or other material placed to stop transfer of water or water vapor.

Water-Reducing Admixture. Material placed in concrete mix to increase slump or maintain workability with a reduced amount of water.

Water Repellent. 1. A finish that is resistant but not impervious to penetration by water. 2. A liquid that penetrates wood which, after drying, materially retards changes in moisture content and in dimensions without adversely altering the desirable properties of wood.

Water-Repellent Paper. Gypsumboard paper surfacing which has been formulated or treated to resist water penetration.

Water Resistant Core. A special gypsumboard core formulation with additives to reduce water absorption; water resistant gypsum backing board is recommended for use as a base for ceramic tile in bathrooms and other wet areas.

Water Resistant Gypsum Backing Board. A gypsumboard designed for use on walls as a base for the application of ceramic or plastic tile.

Water Resistant Gypsumboard. A gypsumboard designed for use in locations where it may be exposed to occasional dampness; plasterboard that has had a chemical treatment to make it resistant to moisture, but not necessarily waterproof.

Water Retentivity. That property of a mortar which prevents the rapid loss of water to masonry units of high suction; it prevents bleeding or water gain when mortar is in contact with relatively impervious units.

Water Service Pipe. The pipe from the water main or other source of water supply to the building serviced.

Water Slide. A sloping trough down which water is carried by gravity.

Water Softener. A device attached to a water system to remove unwanted minerals and substances.

Water Spotting. Spotty changes in the color or gloss of a paint film; may be caused by various factors, such as emulsification or the solution of water soluble components.

Water Stain. A colored dye that is soluble in water.

Waterstop. See Waterbar.

Water-Struck Brick. See Soft Mud Brick.

Water Supply System. The water supply system of a building is composed of the water service pipe, the water distributing pipes and the various connecting pipes, control valves and fittings in or on private property.

Water Table. 1. The level at which the pressure of water in the soil is equal to atmospheric pressure; effectively, the level to which ground water will fill an excavation. 2. The finish at the bottom of a house which carries water away from the foundation. 3. A projection on the bottom of an exterior wall to prevent rain or water from seeping through to the wall below.

Water-Thinned Paint. A paint whose thinner is mainly water; the binder may be a material that (1) requires water for setting, e.g., portland cement; (2) which is soluble in water, e.g., casein; (3) which is emulsifiable in water, e.g., flat wall paint binders.

Watertight. So constructed that moisture will not enter the enclosing case.

Watertight Manhole. A cover for a vertical access shaft that prevents the elements from coming in.

Water-Tube Level. A leveling device consisting of a water-filled tube with a transparent section at each end; also called Water Level.

Water Valve. A device to regulate the flow of water in a pipe or other passage.

Water Vapor Transmission (WVT). See Permeability.

Water Vapor. Moisture existing as gas in the air.

Water White. Transparent and colorless like water.

Watt. 1. A unit of electrical power. 2. The absolute meter-kilogram-second unit of power equal to the work done at the rate of one absolute joule per second. 3. The rate of work represented by a current of one ampere under a pressure of one volt and taken as the standard in the U.S. of 1/746 horsepower.

Watt-Hour. A unit of work or energy equivalent to the power of one watt operating for one hour.

Wattmeter. An instrument for measuring electric power in watts.

Wave. Any disturbance that advances through a medium with a speed that is completely determined by properties of that medium, such as sound or light.

Wave Length. 1. The distance between successive similar points on two wave cycles. 2. In color, the computed distance between vibrations of light that produce visible color sensation on the eye; in the visible spectrum, red-orange has the longest wave length; violet the shortest; wave lengths are measured in millimicrons; wave lengths shorter than violet are called ultraviolet; wave lengths longer than red-orange are called infrared.

Wavy-Grained Wood. Wood in which the fibers collectively take the form of waves or undulations.

Wax. 1. A fatty material obtained from the honeycombs of bees or from similar plant, animal, or mineral substances; used for providing an attractive, protective coating, as for wood; waxes may be used by themselves or combined with other ingredients to make certain polishes, paints, varnishes, and paint removers. 2. Ingredient in many lubricating oils which may separate from the oil if cooled enough.

Wax, Floor. See Floor Wax.

WC. Water closet.

WCLIB. West Coast Lumber Inspection Bureau.

Wearing Course. A topping or surface treatment to increase the resistance of a pavement or slab to abrasion.

Weatherability. Capability of withstanding the weathering process.

Weather Cock. A weather vane in the shape of a cock.

Weather Delay. Lack of construction progress caused by inclement weather.

Weathered. In masonry, stonework which has been cut with sloped surfaces so it will shed water from rain or snow.

Weathered Joint. A mortar joint finished in a sloping profile that tends to shed water to the outside of the wall.

Weatherhead. A weather resistant fitting for feeding aerial electrical conductors into a building.

Weathering. 1. Changes in color, texture, strength, chemical composition or other properties of a natural or artificial material due to the action of the weather. 2. The mechanical or chemical disintegration and discoloration of a wood surface resulting from exposure to light, action of dust and sand carried by winds, alternate shrinking and swelling brought about by changes in the weather or a combination of these causes; weathering does not include decay.

Weathering Steel. A steel alloy that forms a tenacious, self-protecting rust layer when exposed to the atmosphere.

Weather Joint. A masonry joint where the mortar has been formed at a 45 degree angle, sloping down and out, exposing approximately 1/4 to 1/2 inch of the brick above, for the purpose of shedding water.

Weatherometer. A testing device intended to simulate atmospheric weathering.

Weatherproof. So constructed that exposure to the weather will not cause damage or loss of function.

Weatherproof Box. An electrical outlet or switch box which has been manufactured to withstand the outside elements.

Weatherproof Cover. A protective layer constructed so that exposure to the weather will not interfere with successful operation.

Weatherproof Duct. An enclosure for wires or cables constructed so that exposure to weather will not interfere with successful operation.

Weatherproof Insulation. A material used to reduce heat transfer, constructed so that exposure to the weather will not interfere with its successful operation.

Weatherproof Switch. A device designed to be used outdoors that opens, closes or changes the connection of an electric circuit and is weatherproofed against moisture.

Weather Seal. A flanged channel installed on the edges of an exterior door.

Weatherstrip. A ribbon of resilient material used to reduce air infiltration through the crack around a sash or door.

Weather-stripping. The process of reducing air or rain infiltration by covering joints of doors or windows with strips of resilient material.

Weather Vane. A vane; a moveable device for showing the direction of the wind.

Weave Bead. A type of weld bead made with transverse oscillation.

Weaving. Process of forming carpet on a loom by interlacing the warp and weft yarns.

Web. 1. The vertical plate connecting the top and bottom flanges of a metal beam. 2. An interior solid portion of a hollow concrete block.

Web Belt. Commonly a used military type belt adapted to carry hand tools at the jobsite.

Web Members. The vertical and angular members between the top and bottom chords of a truss.

Wedge. A piece of material, such as wood or metal, tapering to a sharp edge that is driven between two objects or parts of an object to secure or separate them.

Weed Control. The act or process of spraying chemicals or placing powders to control the spread of weeds.

Weep Hole. 1. A small opening, the purpose of which is to permit drainage of water that accumulates inside a building component. 2. Openings placed in mortar joints of facing material at the level of flashing to permit the escape of moisture. 3. Openings in retaining walls to permit water to escape.

Weep Screed. A plaster screed at the bottom of an exterior stucco wall that is designed to allow moisture to seep out.

Weeping Joints. A masonry joint treatment in which mortar extruding from the joint in laying is not cut off, but is allowed to harden; gives informal rustic appearance but difficult to waterproof.

Weft. Backing yarns which run across the width of the carpet; in woven carpets, the weft shot (filling) yarns and the warp chain (binder) yarns interlock and bind the pile tufts to the backing; in tufted carpets, pile yarns which run across the carpet are also considered weft yarns.

Weight. The force experienced by a body as a result of the earth's gravitation.

Weir. 1. A dam in a stream over which water flows; see Crest. 2. A dam in a stream to raise the water level or to divert the stream. 3. An orifice through which the liquid only partially flows.

Weir of a Trap. The highest portion of the inside channel through a plumbing trap; compare with Dip of a Trap; the depth of the water seal is the distance from the dip to the crown weir.

Weld. 1. A joint between two pieces of metal formed by fusing the pieces together, usually with the aid of additional metal melted from a rod or electrode. 2. To join two pieces of metal together by heating until fusion of material either with or without filler metal.

Weldability. The capacity of a metal to be welded under the fabrication conditions imposed into a specific, suitably designed structure and to perform satisfactorily in the intended service.

Weld Bead. A weld deposit resulting from a pass.

Weld Crack. A crack in weld metal.

Weld, Destructive Test. Methods to determine existence and extent of defects and discontinuities in welds which affect capabilities of weld and may require repairs after testing.

Welded Joint. A union of two or more members produced by the application of a welding process.

Welded Pipe. Piping where connections and fittings are welded.

Welded Railing. Railing sections with the components fastened with welds.

Welded Truss. Trusses with components fastened together with welds.

Welded Wire Fabric. A series of longitudinal and transverse steel wires arranged substantially at right angles to each other and welded together at the intersections, in sheets or rolls, used to reinforce mortar and concrete.

Welder. One who is capable of performing manual or semiautomatic welding operations based on training, experience, testing, or certification, or any combination of these.

Welding. Fusing metallic parts by heating and allowing the metals to flow together.

Welding Electrodes and Rod. The electrode and rod are the components of the welding circuit through which current is conducted between the electrode holder and the arc.

Welding Goggles. Goggles with tinted lenses, used during welding or oxygen cutting, which protect the eyes from harmful radiation and flying particles.

Welding Inspector. One who is capable of inspection of welds based on training, experience, testing, or certification, or any combination of these.

Welding Leads. The work lead and electrode lead of an arc-welding circuit.

Welding Process. A metal-joining process wherein coalescence is produced by heating to suitable temperatures, with or without the application of pressure, and with or without the use of filler metal.

Welding Rod. Filler metal, in wire or rod form, used in gas welding and brazing processes, and in those arc-welding processes wherein the electrode does not furnish the filler metal.

Welding Technique. The details of a manual, machine or semi-automatic welding operation which within the limitations of the prescribed joint welding procedure are controlled by the welder.

Welding Test. The act or process of testing the strength of a weld.

Welding Tip. A welding-torch tip designed for welding.

Welding Torch. A device used in gas welding or torch brazing for mixing and controlling the flow of gases.

Welding Transformer. A transformer used to supply current for welding.

Weld Inspection. Methods to determine existence and extent of defects and discontinuities in welds.

Weld, Liquid Penetrant Examined. Nondestructive method of inspection to determine existence and extent of discontinuities that are open to surface in weld being inspected; indications are made visible through use of dye or fluorescent chemical in liquid employed as inspection medium.

Weldment Connection. The assembling together of pieces by welding to create a unit.

Weld, Nondestructive Examined. Nondestructive method of inspection to determine existence and extent of discontinuities in welds; finely divided magnetic particles applied to magnetized part are attracted to any magnetic leakage created by discontinuities in mass of sample; methods to determine existence and extent of defects and discontinuities in welds which do not affect capabilities of weld.

Weldor. See Welder.

Weld Plate. A steel plate anchored into the surface of concrete, to which another steel element can be welded.

Weld, Radiographic Examined. Nondestructive method of inspection to determine existence and extent of discontinuities in welds; X rays are used to penetrate the weld to produce radiograph to detect defect.

Weld, Shielded Metal-Arc. See Shielded Metal-Arc Weld.

Weld Test. Welding examination or inspection; the loading of welds to determine load capacity of welds is not normal practice.

Weld, Ultrasonic Examined. Nondestructive method of inspection to determine existence and extent of discontinuities in welds; ultrasonic waves (frequency of mechanical vibrations above 20,000 vibrations per second) are used to penetrate weld to detect defects.

Weld, Visual Examined. Nondestructive method of inspection to determine existence and extent of discontinuities in welds; exposed weld surfaces are visually inspected (without auxiliary equipment) for evidence of existence and extent of discontinuities in welds.

Weld X-Ray. To examine, treat, or photograph the connection of surfaces that have been welded together.

Well. 1. A pit or hole sunk into the earth to reach a supply of water. 2. An open space extending vertically through floors of a structure.

Well Graded Aggregate. Aggregate having a particle size distribution which will produce maximum density of concrete or asphalt, that is, minimum void space.

Wellpoint. A perforated pipe surrounded by sand to permit the pumping of ground water.

Wellpoint System. A series of vertical pipes in the ground connected to a header and pump to drain marshy areas or to control ground seepage.

Western Frame. See Platform Frame.

Wet and Dry Bulb Hygrometer. See Hygrometer.

Wet Areas. Interior or exterior tiled areas subject to periodic or constant wetting; examples are showers; sunken tubs; pools; exterior walls; roofs; exterior paving, and interior floors.

Wet-bulb Temperature. The temperature indicated by the wet-bulb thermometer of a psychrometer.

Wet Cell Battery. Cell or connected group of cells that converts chemical energy into electrical energy by reversible chemical reactions.

Wet-Edge. Fluid paint boundary.

Wet Edge Time. The length of time before a stretch of paint sets up without showing lap marks when the painter applies the next stretch.

Wet Film Gauge. In painting, a device for measuring wet film thickness.

Wet Film Thickness. Thickness of liquid film of paint immediately after application.

Wet Heat. Heating system using hot water heat or steam heat; see Hydronic.

Wet Location Fluorescent. A watertight fluorescent fixture that is sealed to protect against moisture.

Wet Location. A location subject to saturation with water or other liquids, such as exposed to weather, washrooms in garages and like location; installations underground or on concrete slabs or masonry in direct contact with the earth are wet locations.

Wet Mixing Period. The interval of time between the beginning of application of asphalt material and the opening of the mixer gate.

Wet Niche. Underwater swimming pool light with a water cooled sealed beam unit.

Wet Sand. To smooth a finished joint of sheetrock with a coarse wet sponge. A preferred method to reduce dust created in the dry sanding method.

Wet Spray. Paint spraying so that surface is covered with paint that has not started to dry.

Wet Sprinkler System. A sprinkler system that is filled with water at design pressure for immediate use upon activation.

Wet Standpipe. A fire fighting pipeline in a building that is always full of water under pressure and ready for use.

Wettability. A condition of a surface that determines how fast a liquid will wet and spread on the surface or if it will be repelled and not spread on the surface.

Wetting. The process in which a liquid spontaneously adheres to and spreads on a solid surface; the more viscous a fluid, and the higher its surface tension, the more difficult it is for the liquid to wet a material; certain additives, for example, water softeners, reduce surface tension, or viscosity and improve wetting properties, allowing the material to better flow.

Wetting Agent. A substance capable of lowering the surface tension of liquids, facilitating the wetting of solid surfaces and permitting the penetration of liquids into capillaries.

Wetting Oils. Products used to promote adhesion of applied coatings when all mill scale and rust cannot be removed.

Wet Vent. A pipe that acts as a drain or waste pipe and also as a vent when not carrying any liquids; this wet vent usually serves as a waste pipe for the fixture it is closest to; when this fixture is not in use, but other fixtures farther down the line are in use the wet vent acts as a vent pipe for these fixtures farther down the line.

WF Beam. Wide Flange Beam.

WFI. Wood Flooring Institute.

WH. 1. Wall Hydrant. 2. Water Heater.

Wharf. A structure that provides berthing space for vessels, to facilitate loading and discharge of cargo.

Wheel Barrow. A small vehicle with handles and one or more wheels used for carrying small loads.

Wheel Chair Partition. A dividing wall in a bathroom or bathing room which forms the perimeter of a private area that has been made accessible to the disabled.

Wheelchair. A chair mounted on wheels to be propelled by its occupant manually or with the aid of electric power, of a size and configuration conforming to the recognized standard models of the trade.

Wheelchair Occupant. A person who, due to a physical impairment or disability, utilizes a wheelchair for mobility; also called Wheelchair User.

Wheelchair User. See Wheelchair Occupant.

Wheeled Extinguisher. A fire extinguisher mounted on a wheeled cart that can be pushed or pulled by a person.

Wheel Stop. A concrete or wood bumper, approximately 4 to 6 feet (1.20 to 1.80 meters) long, at the end of a parking space to stop the car.

Whiskering. See Efflorescence.

White Blast. Blast cleaning to white metal.

White Cedar Shingle. A light colored weather-resistant cedar wood used for roofing and siding.

White Cement. Cement made from materials with low iron content to produce mortar or concrete that is white in color.

White Coat. A term denoting a gauged lime putty trowel finish.

White Lead. 1. Basic carbonate white lead is a compound of lead, carbon dioxide and water; lead is melted and cast into disks or buckles, which are about six inches in diameter; the buckles are placed in porcelain pots containing dilute acetic acid; the pots are covered with boards and layers of tanbark; heat and carbonic acid generated by fermentation of the tanbark, with the acid vapors, combine to transform the lead into basic carbonate white lead. 2. Basic sulphate white pigment is obtained from lead sulphide ore by a process of fuming or burning. 3. Carter process white lead is made by blowing melted lead in fine granules by a jet of air or superheated steam; the powdered lead is placed in revolving drums or cylinders and subjected to the action of air and carbon dioxide gas from burning coal.

White Metal. Any of several lead-base or tin-base alloys, such as babbitt metal, used for bearings and fusible plugs.

White-Rot. In wood, any decay or rot attacking both the cellulose and the lignin, producing a generally whitish residue that may be spongy or stringy rot, or occur as pocket rot.

Whitewash. A solution of quicklime or of whiting and size for whitening walls.

Whiting. Calcium carbonate, limestone, or chalk in pigment form; used extensively for making putty and as an extender in paints.

WHO. World Health Organization.

WI. Wrought Iron.

Wicking. The absorption of water by capillary action into the core of gypsumboard.

Wide Flange Section. Any of a wide range of steel sections rolled in the shape of a letter T or H.

Wide-Selvage Asphalt Roll Roofing Surfaced with Mineral Granules. See Nineteen-Inch Selvage.

Wide Stile Door. Wider than normal vertical members forming the outside framework of a door.

Width and Length. In a swimming pool, the actual water dimension taken from wall to wall at the maximum operating water level.

Wilton Carpet. A fabric woven on a loom controlled by a Jacquard pattern device, which raises one of from 2 to 6 surface yarns over a bladed pile wire that is then withdrawn to cut the tufts and give a plushlike face; the other yarns run dormant through the center and back of the fabric in the warp direction.

Winch. A hoist or windlass for lifting or moving heavy loads.

Wind. The surface of a board when twisted (winding) or when resting upon two diagonally opposite corners, if laid upon a perfectly flat surface; wind rhymes with kind.

Wind Brace. A diagonal structural member whose function is to stabilize a frame against lateral forces.

Windchill. The still-air temperature that has the same cooling effect on exposed human flesh as a given combination of temperature and wind speed; also called Chill Factor, Windchill Factor, or Windchill Index.

Windchill Factor. See Windchill.

Windchill Index. See Windchill.

Wind Drift. Horizontal deflection of a frame caused by wind forces.

Winder. A stair tread that is wider at one end than at the other, located at a turn in the stairway.

Wind farm. An area of land with a cluster of wind turbines for producing electrical power.

Wind Load. A load on a building caused by wind pressure.

Window. An opening in the wall of a building or structure for the admission of light and air, closed by casements or sashes containing glass panes.

Window Blind. An adjustable window covering.

Window Box. A box on an outside window sill for growing flowers.

Window Covering. Any window treatment such as curtains, draperies, or blinds, for regulating light, for decoration, or for privacy.

Window, Drive-Up. See Drive-Up Window.

Window Frame. The structure which holds a window assembly in place.

Window Guard, Diamond Mesh. See Diamond Mesh Window Guard.

Window Guard Lock. Tamperproof hasp and padlock for window guards.

Window Guard, Steel Bar Grille. Guard fabricated from a steel bar grille.

Window Guard, Woven Wire. Guard fabricated of woven wire.

Window Head. The assembly of parts at the top of a window frame including the frame, stops, casing, shims, and flashing.

Window Header. A horizontal structural member placed across the top of a window opening to support the load above.

Window Pane. A pane of glass in a window.

Window Screen. A fine mesh screen fitted in a window frame to exclude insects.

Window Seat. A seat built into a window recess.

Window, Service. See Service Window.

Window Shade. Any adjustable window covering to regulate light or provide privacy.

Window Sill. The horizontal member at the bottom of a window.

Window Stool. Wood, ceramic tile, or masonry installed on the plate at a window sill on the inside of the window, fitted against the bottom rail of the lower sash.

Window Treatment. The addition of hanging fabrics, curtains, or blinds to the interior of a window.

Window Unit. Air conditioner which is placed in a window.

Windowwall. The opening in a wall surface which contains a window assembly or wall of assemblies.

Window Well. Recess located at or below grade to allow for natural light to reach a ground level or basement window, often created by use of corrugated metal in half-round shape.

Wind Pressure. The horizontal live load on the side of a structure imposed by the wind; it increases in magnitude at higher elevations and causes a suction on the leeward side of buildings and roofs.

Wind Restraint System. The collection of structural elements which provide restraint of the seismic-isolated structure for wind loads; the wind-restraint system may be either an integral part of isolator units or may be a separate device.

Wind Shelf. In a fireplace, the concave ledge just back of the damper at the bottom of the smoke chamber; its function is to direct the cold air downdraft in the flue to join the hot air in rising in the flue.

Wind Tunnel. A tunnel-like passage through which wind is blown at a known velocity past a model of an object to determine the effect of wind forces on the object.

Wind Uplift. Upward forces on a structure caused by negative aerodynamic pressures that result from certain wind conditions.

Wiping Stain. See Pigment Oil Stain.

Wire. A metal drawn out into the form of a thread or thin flexible rod, used for fencing, binding, or to conduct an electrical current.

Wire, Aluminum. See Aluminum Wire.

Wire Brush. 1. A hand cleaning tool comprised of wires for bristles. 2. The act of cleaning a surface with a wire brush, including power brushes.

Wire, Chicken. See Chicken Wire.

Wire Cloth Lath. A plaster reinforcement of wire not lighter than No. 19 gauge, 2-1/2 meshes per inch and coated with zinc or rust-inhibitive paint; not to be used as reinforcement of exterior portland cement plaster.

Wire-Cut Brick. Brick formed by forcing plastic clay through a rectangular opening designed for the purpose, and shaping of clay into bars; before burning, wires pressed through the plastic mass cut the bars into brick units.

Wire Edge Joint. In wallpaper, a joint made by trimming both selvedges and lapping one edge slightly over the other.

Wire Glass. Glass in which a wire mesh was embedded during manufacture.

Wire Guard. Flexible strands of metal that have been manufactured into a unit to act as an enclosure around moving parts of machinery, around an excavation, equipment, or materials to prevent injury to the operator.

Wire Hanger. A wire that supports or connects material.

Wire Hook Hanger. Flexible strand of metal in the shape of a hook to hold a construction member in place.

Wire Mesh. A series of longitudinal and transverse wires arranged substantially at right angles to each other sheets or rolls, used to reinforce mortar and concrete.

Wire Mesh Partition. Dividing wall constructed of metal framing and wire mesh.

Wire Mesh Partition Adjustable Cap. Top capping channel that is tightly fitted over top of entire partition.

Wire Mesh Partition Adjustable Floor Shoe. Adjustable floor socket to permit level installation of wire mesh partition channel.

Wire Nut. A device for connecting electrical wire conductors together by twisting.

Wires. In carpet making, metal strips inserted in the weaving shed under the surface yarns to form loops when the yarns are bound by the weft shuttle in the Velvet and Wilton weaves; a round wire will withdraw, leaving uncut pile loops for round-wire Velvet and Wiltons, while bladed knife-ends on flat wires will cut these loops, forming a plushlike surface of tufts; the term has come to indicate the number of rows of tufts per inch of warp; thus 11 wires means 11 rows of tufts to each inch of length.

Wire, Stranded. See Stranded Wire.

Wireway. A sheet metal trough with hinge or removable cover to carry several electrical cables.

Wiring, Flexible. See Flexible Wiring.

With Framing. Of gypsumboard application, see Parallel Application.

Withdrawing Room. Drawing Room.

Withe. 1. A single tier of brick or stone in a wall. 2. A masonry separation or partition between flues in a chimney with more than one flue.

Withering. Withering or loss of gloss is often caused by varnishing open-pore woods without filling pores, use of improper undercoating and applying topcoat before undercoat dried.

Witness. A person who observed events and is called to testify concerning those events at a hearing.

Wobble Friction. In prestressed concrete, friction caused by unintended deviation of prestressing sheath or duct from its specified profile.

Wobble Plate-Swash Plate. Type of compressor designed to compress gas, with piston motion parallel to crankshaft.

Wolfram. See Tungsten.

Wolmanized Lumber. Lumber treated with Wolman salts to resist decay.

Wood. The hard fibrous material that makes up most of the trunk and branches of trees.

Wood Alcohol. Poisonous alcohol obtained by destructive distillation of wood.

Wood Anchor. A bolt or fastening device which attaches wood to wood or wood to another material.

Wood and Plastics. A category of the CSI Masterformat which is represented in Division 6 of the format. Commonly called just Wood or Carpentry.

Wood Backing. See Backing, 2.

Wood Batten. Wood strips covering vertical joints on boards used as exterior siding.

Wood Beam. Horizontal wood structural member that supports uniform and concentrated loads.

Wood Blocking. Small pieces of wood used to secure, join, or reinforce members, or to fill spaces between members.

Wood Board Roof Sheathing. Wood board material placed diagonally and secured to exterior side of roof rafters or trusses used to create rigidity in building superstructure and serve as base to receive roofing.

Wood Board Subflooring. Wood board material placed diagonally and secured to top side of floor joists used to create rigidity in building superstructure and serve as base to receive flooring.

Wood Board Wall Sheathing. Wood board material placed diagonally and secured to exterior side of exterior wall studs used to create rigidity in building superstructure and serve as base to receive siding or veneer construction.

Wood Bottom Plate. 1. In Western or Platform Framing construction, a horizontal wood lumber member which supports wall studs and ties them together; rests on the floor sheathing. 2. In Balloon Framing construction, a flat horizontal member, also called a mudsill, that supports the vertical wall studs and posts.

Wood Bridging. Diagonal or longitudinal wood members used to keep wood joist members properly spaced, in lateral position, vertically plumb, and to distribute load.

Wood Buck. Wood frame typically built into concrete or masonry wall to accommodate finish door frame.

Wood Bumper. Wood component used to absorb impact and prevent damage to other surfaces, such as at a loading dock.

Wood Cant Strip. Sloped wood strip used at perimeter of roofing membrane to transition membrane from horizontal to vertical surface.

Wood Cap. Wood member used on top of an assembly to provide termination or finish.

Wood Carriage. Sloping beam installed between stringers to support steps of wood stair.

Wood Chip Mulch. Wood chips, spread on the ground to prevent erosion, control weeds, minimize evaporation, and improve the soil.

Wood Chisel. A sharp ended hand tool used to carve special shapes into wood.

Wood Column. Vertical wood structural member, usually supporting a beam.

Wood Cornice. Horizontal wood molding that may be combination of several shaped pieces.

Wood Decking. Plywood, lumber, or glued laminated member placed over roof or floor structural members for structural rigidity of building frame and to provide a surface for traffic or substrate for roofing or flooring system.

Wood Diagonal Bracing. Diagonal wood member used to prevent buckling or rotation of wood studs.

Wooden Brick. Piece of seasoned wood, made the size of a brick, and laid where it is necessary to provide a nailing space in masonry walls.

Wood Failure. The rupturing of wood fibers in strength tests of bonded joints usually expressed as the percentage of the total area involved which shows such failure.

Wood Fascia. Flat vertical wood member of cornice, eaves, or gable or other finish, generally that part of the assembly to which the gutter is secured.

Wood Fiber. 1. A wood cell, comparatively long (1/25 or less to 1/3 inch) (1 to 8 millimeters), narrow, tapering, and closed at both ends. 2. Ground or shredded, non-staining wood used as an aggregate with gypsum plaster.

Wood-Fibered Plaster. Neat gypsum basecoat plaster containing wood fiber as an aggregate and designed for use either with or without addition of other aggregates.

Wood Filler. A heavily pigmented preparation used for filling and leveling off the pores in open-grained woods.

Wood Finish Concrete. The act or process of using a wood float to smooth irregularities left in curing concrete; work the surface or compact the concrete.

Wood Float. A flat wooden trowel used for floating mortar, normally used just before the pure coat.

Wood Floor Joist. Horizontal structural member of a framed floor.

Wood Floor Removal. The act or process of tearing up old wood floors.

Wood Flooring. Floor coverings consisting of dressed and finished boards.

Wood Flour. Wood reduced to finely divided particles, approximately those of cereal flours in size, appearance, and texture, and passing a 40-100 mesh screen.

Wood Frame. Floors, roofs, exterior and bearing walls of a building or structure constructed with wood.

Wood Furring. Strips of wood applied to surfaces (usually concrete or masonry) to provide a planar surface and to provide a fastening base for finish material.

Wood Girder. Large horizontal wood beam which supports concentrated loads at isolated points along its length such as the support of joists or rafters.

Wood Ground. Narrow wood strips used around openings and at the perimeter to provide a guide for strike off of plaster to provide a straight and level or plumb line for plaster.

Wood Header. Wood member placed across joist ends or at openings in a wall to support joists or studs at openings in a framing system.

Wood Joist. Horizontal framing member of a floor, ceiling or flat roof.

Wood Lath. Strips of wood, 1-1/4 inches wide by 3/8 inch thick, formerly used as a plaster base; now entirely replaced by gypsum lath.

Wood Lintel. Wood header over openings in wood frame wall construction.

Wood Nailer. Strip of wood attached to steel or concrete to facilitate making nailed connections.

Wood Pile. A long, slender wooden pole driven into the ground to act as a foundation; a member embedded into the ground that supports vertical loads.

Wood Plate. Horizontal wood lumber member on top or bottom of wall studs which ties them together and supports studs, joists, or rafters.

Wood Pole. A long piece of wood used to carry utility lines.

Wood Railing Newel. Principal wood post at foot of stairway or central support of a winding flight of stairs.

Wood Railing Post. Large vertical wood member to support railing.

Wood Rays. Strips of cells extending radially within a tree and varying in height from a few cells in some species to 4 or more inches in oak; the rays serve primarily to store food and transport it horizontally in the tree; on quartersawed oak, the rays form a conspicuous figure, sometimes referred to as Flecks.

Wood Ridge Rafter. Horizontal wood supporting member at top of sloping roof immediately beneath sheathing.

Wood Riser. The vertical wood board under the tread in a stairway system.

Wood Roof Cricket. Relatively small elevated area of wood roof constructed to divert water around chimney, curb, or other projection.

Wood Roof Curb. Wood member elevated above plane of roof surface used for mounting of equipment or other elements.

Wood Roof Decking. Plywood, lumber, or glued laminated member placed over roof structural members for structural rigidity of building frame and to proved surface for traffic or substrate for roofing system.

Wood Roof Edge Strip. Wood strip (usually in plane of roof insulation) at perimeter of roof secured to structural roof deck used for securing roofing membrane.

Wood Roof Nailer. Wood strip (usually in plane of roof insulation) secured to structural roof deck used for securing roofing membrane.

Wood Saddle. Short horizontal wood member set on top of wood column to serve as seat for a girder.

Wood Saw. See Handsaw.

Wood Screw. A screw for fastening objects in wood.

Wood Shake. A hand-split shingle.

Wood Shingle. Factory cut and shaped roof covering of wood (usually of cedar), cut into modular lengths, widths, and rectangular profile.

Wood Siding. Wood used as exterior surface or cladding for exterior framed wall to provide protection from the elements.

Wood Sill Plate. Horizontal wood lumber member on bottom of wall studs which ties them together and rests on concrete or masonry; horizontal timbers of a house which either rest upon the masonry foundations or, in the absence of such, form the foundation.

Wood Sleeper. Wood member laid on concrete floor to support and receive fastening of wood subfloor or finish flooring.

Wood Stain. Finish for wood containing a dye or pigment; stain sinks into fibers of the wood to a certain extent while paint and lacquer ordinarily do not penetrate the wood.

Wood Stair Framing. Wood structural members supporting stairs or stair openings.

Wood Structural Panel. A structural panel product composed primarily of wood; wood structural panels include all-veneer plywood, composite panels containing a combination of veneer and wood-based material, and mat-formed panels such as oriented strand board and waferboard.

Wood Substance. The solid material of which wood is composed.

Wood Threshold. Wood strip fastened to floor beneath door.

Wood Treatment. The act or process of applying a variety of stains or chemicals to retard fire, decay, insect damage or deterioration, due to the elements.

Wood Trim. Wood millwork, primarily moldings, used to finish off and cover joints and openings.

Wood Truss. A structural component formed of wood members in a triangular arrangement, often used to support a roof.

Wood Truss Joist. Joists of rigid open framework construction with top and bottom chords, fabricated of wood web and chord members or combination wood chord members with metal web members.

Wood Truss Rafter. Truss where chord members also serve as rafters and ceiling joists.

Wood Wool. Long, curly, slender strands of wood used as an aggregate component for some particleboards.

Woodwork, Architectural. A higher than average quality feature of finish work using wood for ornamental design.

Wool. 1. Fine soft wavy hair from the fleece of sheep, goats, and other animals. 2. A yarn produced from this hair. 3. Fabric made from it. 4. Any of various wool-like substances like steel wool, lead wool, or mineral wool.

Woolen Yarn. Soft, bulky yarn spun from both long and short wool fibers which are not combed straight but lie in all directions so they will interlock to produce a felt-like texture.

Wool, Mineral. Any of various lightweight vitreous fibrous materials used for heat and sound insulation.

Workability. 1.The ease with which material can be worked or smoothly cut and shaped with hand or machine tools. 2.In masonry, the texture of mortar such that it behaves properly under the trowel. 3. In painting, the texture and consistency of the paint such that it spreads properly. 4. In woodworking, the degree of ease and smoothness of cut obtainable with hand or machine tools. 5. The property of freshly mixed concrete or mortar which determines the ease and homogeneity with which it can be mixed, placed, compacted, and finished. 6. A property of plaster mortar closely related to plasticity which determines the ease and speed with which the mortar can be applied and finished. 7. The ease with which paving mixtures may be placed and compacted.

Workbench. Table at which work is accomplished.

Workers Compensation. A system established by statute under which employers are responsible for medical expenses and disabilities of workers injured while on the job; compensation is payable even if the employer is not at fault and even if the carelessness of the worker contributed to the injury, but the employer is not necessarily liable for damages for pain and suffering.

Working Capital. The amount of capital available for current use in the operations of a business measured by the excess current assets (cash and assets readily converted into cash) after current liabilities have been subtracted.

Working Day. A day on which construction work may be done, eliminating Saturdays, Sundays, and Holidays.

Working Drawings. Drawings of the project that are used in the construction of structure, they are part of the contract documents.

Working Life. The period of time during which an adhesive, sealant, or other material, after mixing with catalyst, solvent, or other compounding ingredients, remains suitable for use; also called Pot Life.

Working Properties. The properties of an adhesive that affect or dictate the manner of application to the adherends to be bonded and the assembly of the joint before pressure application, such as viscosity, pot life, assembly time, and setting time.

Working Stress. The maximum permissible stress used in the working stress design of a member.

Working Stress Design. A design theory that is used in the design of concrete and masonry members; safety is provided by limiting allowable stresses.

Work Load. The electric conductor between the source of arc-welding current and the work.

Workout. A procedure whereby a borrower negotiates a restructuring of a mortgage or trust deed in lieu of a foreclosure.

Work Product Exclusion. A provision in a liability insurance policy that withholds coverage for damages to work performed by or on behalf of the insured.

Work Station. An area defined by equipment or work surfaces intended for use by employees only, and usually for one or a small number of employees at a time; examples include ticket booths, the employee side of grocery store checkstands, the bartender area behind a bar, the employee side of snack bars, sales counters and public counters, guardhouses, toll booths, kiosk vending stands, lifeguard stations, maintenance equipment closets, counter and equipment areas in restaurant kitchens, file rooms, and storage areas.

Work Top. A counter top.

Worm Gear. Gear teeth spirally cut into a shaft meshing with a worm wheel; the worm gear, or worm, and the worm wheel are on perpendicular axes.

Worsted Yarn. Strong, dense yarn made from long staple fibers which are combed to align the fibers and remove extremely short fibers.

Woven Carpet. Carpet made by simultaneously interweaving backing and pile yarns on one of several types of looms from which the carpets derive their names.

Woven Wire Fabric. A plaster reinforcement of zinc-coated wire, not lighter than No. 18 gauge when woven into 1 inch openings, or not lighter than No. 17 gauge when woven into 1-1/2 inch openings; lath may be paper-backed, flat or self-furring.

WP. Water proofing.

Wrench. A hand or power tool, some adjustable, for gripping and turning bolts and nuts.

Wrinkle. 1. A pattern of ridges usually caused by roofing plies not being flat in the bitumen intermopping. 2. See Cockle.

Wrinkled Sheets. Pertaining to ceramic mosaics mounted on paper. Due primarily to rough handling in shipment.

Wrinkle Finish. A varnish or enamel film which exhibits fine wrinkles or ridges. Used extensively as a novelty finish.

Wrinkling. A surface defect resembling the skin of a prune.

Wrist Action. In using a spray gun, swiveling of wrist without arcing forearm.

Wrongful Death. Unlawful homicide, whether by negligence or intent.

Wrought Iron. A tough malleable, relatively soft, form of iron suitable for forging or rolling; not cast.

WWPA. Western Wood Products Association.

Wye. A Y-shaped pipe fitting which joins 3 pipes.

Wythe. 1. Each continuous vertical section or layer of masonry one unit in thickness; also called a Tier. 2. The thickness of masonry separating flues in a chimney.

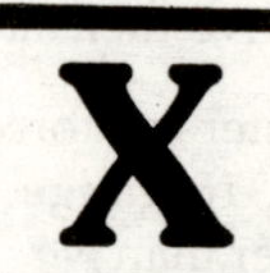

Xanthic. Containing yellow or pertaining to yellow color.

X-Axis. One of the axes in a three-dimensional system, that is not the y- or z-axis.

X Bracing. That form of bracing wherein a pair of diagonal braces cross near mid-length of the bracing members.

Xenon. A heavy colorless odorless inert gaseous element occurring in traces in the atmosphere and used in fluorescent lamps.

Xeric. Requiring only a small amount of moisture, as a xeric habitat.

Xerography. A dry electrostatic document copying process.

Xerothermic. A hot dry environment.

XO, XOO, XOOX. Designations of the arrangement of sliding and fixed panels in sliding glass doors and windows; X indicates a sliding panel; O indicates a fixed panel; see OX, OXO, OXXO.

X-Ray. Electromagnetic radiations of a short wavelength that can penetrate various thicknesses of all solids.

X-Ray, Weld. See Weld X-Ray.

Xylem. The portion of the tree trunk, branches, and roots that lies between the pith and the cambium.

Xylene. A solvent obtained from wood tar, coal tar, or petroleum.

Xylography. The art of making engravings on wood.

Xylol. Xylene.

Xylonite. A thermoplastic material like celluloid, used in model building.

Xyst. In Greek and Roman architecture, a long covered portico for exercise in bad weather.

Y

Yard. 1. A unit of linear measure equal to 3 feet or 36 inches; equal to 0.9144 meters. 2. A square yard, as in yards of carpeting. 3. A cubic yard, as in yards of earth cut or fill. 4. A piece of enclosed ground. 5. The space around a residence; see Curtilage. 6. The space between buildings.

Yard lumber. Run of the mill lumber in a lumber yard, including all sizes and patterns.

Yarn Size. Carpet yarn is measured by the number of yards in length to the ounce of weight in a single ply; two to four plies of this single strand will be twisted to form the best balance for any particular weave and quality of fabric desired; it is necessary to know yarn sizes and plies in order to make fair comparisons.

Y-Axis. One of the axes in a three-dimensional system, that is not the x- or y-axis.

Yellow Fir. Douglas Fir.

Yellowing. Development of yellow color or cast, in whites, on aging.

Yellow Metal. Muntz metal.

Yellow Ochre. An earthy pigment made from impure iron ore, used in paint.

Yellow Ware. A yellow semi-vitreous ware or an earthenware with a colorless clear glaze.

Yield. 1. The ratio of the net income of a real estate or other investment compared to the acquisition and operating costs. 2. The amount of concrete produced by a given combination of materials; the total weight of ingredients divided by the unit weight of the freshly mixed concrete. 3. The cubic feet of concrete produced per sack of cement. 4. The number of product units, such as block, produced per batch of concrete or sack of cement.

Yield Point. The point at which a material deforms with no increase in load; the stress at which a material ceases to deform in a fully elastic manner.

Yield Strength. The specified minimum yield strength or yield point of reinforcement in psi.

Yoke Vent. A pipe connecting upward from a soil or waste stack to a vent stack for the purpose of preventing pressure changes in the stacks.

Young's Modulus. The ratio of the unit stress to the unit strain below the proportional limit.

Y Strainer. A device in the shape of a Y for withholding foreign matter from a flowing liquid or gas.

Z

Z. A numerical coefficient used in the design of earthquake forces and that is dependent upon site location.

Zaffer. An impure oxide of cobalt used as a blue ceramic coloring.

Zax. An implement used for splitting and installing slates; a kind of hatchet with a sharp point on the head for perforating the slate to receive a nail or pin.

Z-Axis. One of the axes in a three-dimensional system, that is not the x- or y-axis.

Z Bar. A Z-shaped steel or sheet metal strip used for light framing or furring.

Zebrawood. The mottled or striped wood of several trees, such as various leguminous African timber trees with pale golden heartwood uniformly striped with dark brown or black.

Zenith. The highest or culminating point.

Zero Ice. Trade name for dry ice. See Dry Ice.

Zero Slump Concrete. A concrete mixed with so little water that it has a slump of zero when tested.

Zeta. A closed or small chamber; a room over a church porch where documents were kept.

Z-Furring Channel. A Z-formed metal channel for mechanically attaching gypsum board and insulation material on masonry walls.

Ziggurat. A stepped pyramid, as in the sacred architecture of Western Asia in antiquity.

Zigzag. Making short and sharp turns; in architecture, especially in the moldings in arched door heads of Romanesque style.

Zinc. 1. A bluish white crystalline metallic element of low to intermediate hardness that is ductile when pure but in the commercial form is brittle at ordinary temperatures and becomes ductile on slight heating. 2. Zinc, as galvanizing, is widely used as a protective coating for iron and steel. 3. Zinc is used extensively as paint pigment.

Zinc Chromate. Metal priming pigment with important rust-inhibitive properties; Zinc Yellow.

Zinc Dust. Finely divided zinc metal, gray in color; used primarily in metal primers.

Zincky. Containing or resembling zinc.

Zinc, Leaded. See Leaded Zinc.

Zincoid. Relating to or resembling zinc

Zinc Oxide. A compound of zinc used as a white pigment in many types of paint.

Zinc Phosphate Coating. Treatment used on steel to improve adhesion of coatings.

Zinc Silicate. Inorganic zinc coating.

Zinc Sulphide. Compound of zinc used as white pigment in paints.

Zinc Yellow. Commercial zinc chromate pigment.

Zippers. See Paper Rollers.

Zip Tape. In gypsumboard bundles, a reinforcement paper strip to facilitate the removal of end bundling tapes.

Zircon Porcelain. A vitreous ceramic whiteware for technical application in which zircon (ZrO_2SiO_2) is the essential crystalline phase.

Zone. 1. In space heating or air conditioning, a specific area in a building that is kept at the same temperature. 2. A defined area of land that is limited in use to some particular purpose, such as residential, commercial, or industrial; see Zoning, 2.

Zoning. 1. Government regulation of the use of privately owned land. 2. The official designation of parts of a municipality or other governmental territory to be used only for certain specified land uses.

Zonolite. A lightweight insulating concrete composed of portland cement, vermiculite aggregate, and water.

Zoophoric Column. A pillar supporting the figure of an animal.

Z Tie, Wall. See Wall Z Tie.